Environmental Science:

A Week By Week Homeschooling Guide

John Turano

INTRODUCTION

As long as there has been life on Earth, there has been the science of ecology. As long as humans have been on Earth, there has been environmental science. Seems strange? Let me explain.

Ecology is the study of how organisms interact with each other and their non-living environment. In prehuman times organisms survived, or not, based on biological principles like population size, available resources, and predator-prey relationships, for example. But they were also at the mercy of environmental changes. Things like volcanic activity, earthquakes, continental drift, and the occasional asteroid hit. The evidence we have now of previous mass extinctions all indicate that they were caused by geologic catastrophes (like getting hit by a huge asteroid) that caused major environmental changes. And life started over pretty much from scratch.

But humans, and human activity, add another dimension to all of this. We have the ability to change the environment. And as long as the human population stayed relatively small, the impact of human activity on the global environment was relatively small. Whatever environmental issues there were, were local in nature. That all changed with the Industrial Revolution. The impact that human activity has had on the global environment since the Industrial Revolution has caused great concern in all parts of the world.

Environmental science integrates the physical and biological sciences to the study of the environment and for the solution of environmental problems. Unlike the study of ecology, which looks at the interactions of organisms to each other and their environment, environmental science includes not only ecology, but physics, chemistry, geology, geography, economics, politics, and more.

This book is intended to be a follow-up to my Basic and Advanced Biology guides. Where the other books provided the foundation for understanding biology, this book is intended for the applications of those foundations to real-world problems.

Let me emphasize that this book is only a **guide** and not a textbook. The intent is to provide a basic framework of topics for as much flexibility as is necessary to cover the topics that you feel are appropriate to your situation. For example, if human population growth is of greater interest than air pollution, I've given you enough flexibility to spend as much time on population growth as possible. The only constraints that you have are the ones you impose on yourself.

There are twenty-five topics separated into six units. The guide is designed to cover a 36-week course of study. But that can be adjusted based on your needs.

UNIT 1 – Environmental Science Basics
UNIT 2 – Biodiversity and Species Interactions
UNIT 3 – Biodiversity and Sustainability
UNIT 4 – Earth's Resources
UNIT 5 - Pollution
UNIT 6 - Environmental Sustainability

Each Unit consists of a number of modules with Units 3 and 4 being the largest, at five modules each. The modules are all laid out the same. There is an overview of the module's topic, followed by the goals for you to meet in the module, then key words, and a progress check.

I recommend that you start with examining the goals and the progress check. They will give you the direction to go in. Then review the key word, and finally read the overview.

At the end of the guide you'll find two lists of web-based resources. One is a list of environmental organizations that cover the entire spectrum of environmental problems. Though they are cause-specific, they do provide a wealth of information about their specific concern.

The other list contains web-based resources that will add depth to the modules. These resources will lead to you research the module topic in more depth. The intent of this list is to not give all the answers, but for you to discover the answers for yourself.

There are two resources worth mentioning. The Habitable Planet (http://www.learner.org/courses/envsci/index.html) is an excellent source of information for almost all of the topics. Before you start with any topic, examine this web site and become familiar with the content. It's time well spent.

Another web-based resource, that isn't on the list, is the National Center For Case Study Teaching in Science (http://

sciencecases.lib.buffalo.edu/cs/). This excellent web site presents case studies in any topic in science for every grade level. If, for example, you search for air pollution, you'll get a number of case studies dealing with air pollution. Most of the time you can download the case study to your computer.

My goal is to provide you with a guide that will allow you to teach your child about environmental science in a manner that is appropriate to his or her life, that will have a lasting effect on how your child sees the living world, and to create a deeper understanding of why the environmental issues we face today are truly critical. The world is changing before our eyes and the more people there are questioning and discussing these issues, the better off the planet will be.

Unit 1 - Environmental Science Basics

Module 1 Causes of Environmental Problems

Module 1 Overview

Unlike ecology, environmental science is a multidisciplinary study whose goal is to look at how we can establish sustainable societies. As such, you must examine all the fields of study, both in the natural sciences and the social sciences, that make up environmental science.

Once you understand the scope of environmental science, you need to understand what the concept of sustainability means. This is the single-most important concept in environmental science. For a good start in understanding sustainability and sustainable development, visit the following two web sites: http://en.wikipedia.org/wiki/Sustainability and http://en.wikipedia.org/wikiSustainable_development.

In spite of Man's scientific and technological achievements, we are totally dependent on the environment for air, water, food, shelter, energy, and everything else we need to stay alive and healthy. What many do not understand is that we are part of, and not apart from, the rest of nature.

A key subfield of environmental science is ecology, the branch of biology that studies how living things interact with their environment and with each other. The focus of ecology is the study of ecosystems; a set of organisms interacting with one another and with their nonliving environment. Do not confuse environmental science and ecology with environmentalism. Environmentalism a social/political movement that protects the earth's life-support systems. Environmentalism is active in the political and ethical arenas rather than in scientific arenas.

There are three major components to sustainability. The first is how we utilize natural capital. Those natural resources available to us that keeps everything on earth alive. Natural resources are the materials, and energy, in nature that are essential and useful to humans. These resources are classified as renewable or non-renewable.

In addition to natural resources are natural services. These are functions of nature, such as purification of air and water or the nutrient cycles which recycle the essential chemicals living things need in order to survive.

The second component of sustainability is recognition that many human activities degrade natural capital by using normally renewable resources faster than nature can renew them.

The third component of sustainability is the search for solutions to problems like the over-use of natural capital. Unfortunately the work of environmental scientists is limited to finding scientific solutions, while the political solutions are left to political processes. And here is where there is conflict.

For example, scientists may argue for the protection of a certain species in an ecosystem which is viewed as a positive thing. On the other hand, business and political interests might conflict with the scientists' views.

And this is the fourth component of sustainability: compromise. The search for solutions always involves conflict and the best outcome is the one where all parties involved are satisfied with the results.

Another aspect of studying the impact of human activity on the environment is a term called ecological footprint. How are our ecological footprints affecting the Earth?

It should be understood that people in developing countries struggle to survive. The individual use of natural resources and the resulting environmental impact is low and is devoted mostly to meeting their basic needs. They are just trying to survive day-by-day. By contrast, many individuals in affluent nations consume large amounts of resources that are way beyond their basic needs.

Supplying people with resources and dealing with the resulting wastes and pollution has a large environmental impact. This is referred to as an ecological footprint - the amount of productive land and water needed to supply people in a particular area with resources, and to absorb and recycle the wastes and produced by this use. What is referred to as the per capita ecological footprint is the average ecological footprint of an individual in a given country or area. The per capita ecological footprint is an estimate of how much of the earth's renewable resources an individual consumes. To see what your ecological footprint is, try taking this quiz at: http://www.myfootprint.org/.

The next concept in this module you need to understand, and one which is understood by almost everyone, is pollution. First you need to know what pollutions is, where it comes from, how it impacts all our lives, what can be done to prevent it and how do we clean it up. Unit 5 of this guide goes into detail about the different forms of pollution.

The last concept is understanding the major causes of environmental problems. There are five accepted causes of environmental problems and they are population growth, wasteful and unsustainable resource use, poverty, ignoring the environmental costs of resource use from the market prices of goods, and attempts to manage nature with insufficient knowledge.

Module 1 Goals

The following is a list of goals that should be met by the end of this module. The goals are very broad

and, sometimes, general in nature. This was done to allow you to decide how much depth you'll want to go into for each goal.

Examining the Progress Check questions will also help focus your attention better on the goals.

1. Define what an environmentally sustainable society is.
2. Explain how environmentally sustainable societies can grow economically.
3. Describe the affects of our ecological footprints on the earth.
4. Define pollution and what can be done about it.
5. Explain why we have environmental problems and their causes.
6. Describe the four principles of sustainability.

Module 1 Key Words

Here's a list of key words for this module:

conservation
ecological footprint
ecology
ecosystem
environmentalism
gross domestic product
natural capital
natural resources
nutrient recycling
organisms
pollution
recycling
renewable resources
solar capital
species
sustainability
sustainable yield

Module 1 Progress Check

To check your understanding of the goals of this module, you should be able to answer the following questions.

1. Distinguish between environmental science, ecology, and environmentalism.
2. Explain the terms natural capital, natural resources, natural services, solar capital, and natural capital degradation.
3. Describe the ultimate goal of an environmentally sustainable society.
4. Distinguish between developed countries and developing countries and describe their key characteristics.
5. What is environmentally sustainable economic development?
6. What is conservation? Distinguish among a renewable resource, nonrenewable resource, and perpetual resource and give an example of each.
7. What is culture and describe three major cultural changes that have occurred since humans arrived on the earth.
8. Why has each of the cultural changes led to more environmental degradation?
9. Identify five basic causes of the environmental problems that we face today.
10. What are the four scientific principles of sustainability and explain how exponential growth affects them.

Module 2 Environmental Systems, Matter, and Energy

Module 2 Overview

This module is a review of what you should already know about what science is and is not. It's truly important that you fully understand scientific processes and how scientists work.

For environmental scientists, having an accurate method of making valid hypotheses based on accurate observations and data collection is critical since their findings can have global impact. For example, the data that Al Gore used in his "An Inconvenient Truth" book was flawed when he claimed that the earth was warming up. Unfortunately his conclusion was missing a lot of data that would have shown that, at that point in time, global warming wasn't as serious as he made it out to be.

Environmental scientists face a number of limitations. First, scientists can disprove things but they cannot prove anything absolutely, because there is always some degree of uncertainty in scientific measurements, observations, and models.

Second, scientists are human and cannot be expected to be totally free of bias about their results and hypotheses. However, bias can be minimized and uncovered, as is the case many times, through peer review.

A third limitation is statistics. For example, there is no way to measure accurately how much soil is eroded annually worldwide. Instead, scientists use statistical sampling and methods to estimate such numbers and indicate important trends.

A fourth problem is that many environmental phenomena involve huge numbers of interacting variables and complex interactions. This makes it too costly to test one variable at a time in controlled experiments.

Finally, scientific processes are limited to understanding the natural world. Scientists do not apply these processes to moral or ethical questions. This is because such questions are about matters for which we cannot collect data. For example, we can use the scientific process to understand the effects of putting too much carbon dioxide into the atmosphere, but this process doesn't tell us whether it's right or wrong to do so.

In addition to understanding the basic concepts of matter, and the chemical nature of biological systems, you also need to understand the nature of complex hydrocarbons, especially in terms of pollution and why they cause damage to the environment. For example, a once commonly used refrigerant gas called CFC (chlorofluorocarbon), known as freon, was found to destroy the ozone layer of the upper atmosphere. Understanding how freon did that allowed scientists to get world governments to ban the sale of freon.

In any basic science course you learn about two very important laws of nature. The first is the Law of Conservation of Energy and Mass. This says that you cannot create nor destroy matter, you can only change its form. The biogeochemical cycles are a good example of what this scientific law is all about.

The second is the Law of Thermodynamics. Like the Law of Conservation, the Law of Thermodynamics illustrates what energy is and how it behaves. Ecology, at its basic form, is really the study of how energy flows through an ecosystem. This is predicated on the fact that all living things need a source of energy to survive. Some organisms produce it, others consume the organisms that store the energy they make. Look at any food web and you get an idea of how energy flows through an ecosystem.

The last important concept in this module is the understanding of systems. A system is a set of components that function and interact in some regular way. The human body, a river, an economy, and the earth are all systems.

Most systems have the following key components: inputs from the environment, flows of matter and energy within the system at certain rates, and outputs to the environment.

Since controlled experimentation is not possible, or desirable, in environmental science, scientists will use computer modeling to study systems and how they interact with one another. Studying climate change relies a great deal on computer modeling to forecast future impacts of our changing climate.

In studying systems, scientists rely on feedback loops. Most systems are affected one way or another by feedback, any process that increase or decreases a change to a system. Feedback loops occur when an output of matter, energy, or information is fed back into the system as an input and leads to changes in that system. A positive feedback loop causes a system to change further in the same direction. A negative feedback loop causes a system to change in the opposite direction from which is it moving.

Module 2 Goals

The following is a list of goals that should be met by the end of this module. The goals are very broad and, sometimes, general in nature. This was done to allow you to decide how much depth you'll want to go into for each goal.

Examining the Progress Check questions will also help focus your attention better on the goals.

1. Define science.
2. Describe what matter is.
3. Explain the ways in which matter can matter change.
4. Define energy and how it can be changed.
5. Define systems and how they respond to change.

Module 2 Key Words

Here's a list of key words for this module:

acidity
atom
atomic theory
chemical change
chemical formula
compounds
data
deductive reasoning
electrons
elements
experiments
feedback loops
hypothesis
inductive reasoning
inorganic compounds
ion
isotopes
law
law of conservation of matter
law of thermodynamics
matter
model
neutron
nucleus
organic compounds
paradigm shift
peer review
pH
physical change
protons
theory

Module 2 Progress Check

To check your understanding of the goals of this module, you should be able to answer the following questions.

1. Describe the steps involved in the scientific process.
2. What is peer review and why is it important?
3. Distinguish between a scientific hypothesis, scientific theory, and scientific law.
4. Distinguish between inductive reasoning and deductive reasoning and give an example of each.
5. What are five limitations of science and environmental science?
6. What is the law of conservation of matter and explain why it is important?
7. What is the law of conservation of energy explain and why it is important?
8. What is the second law of thermodynamics, why is it important and explain why this law means that we can never recycle or reuse high-quality energy.
9. Distinguish between a positive feedback loop and a negative feedback loop in a system, and give an example of each.
10. Explain how human activities can have unintended harmful environmental results.

Unit 2 - Biodiversity and Species Interactions

Module 3 Ecosystems

Module 3 Overview

Ecology is the study of how organisms interact with one another and with their physical environment of matter and energy. Environmental science takes the study of ecology and places it in context with other natural sciences and adds the influences of social science.

So in ecology you consider the cell, both prokaryotic and eukaryotic, and how cells form organisms. From there your can differentiate species and how those species form populations, and how populations form communities, and how communities form ecosystems, and, finally, how ecosystems form biomes. And that's as far as ecologists go. They study all these and how they interact.

Ecologists recognize earth's life-support system consists of four main spherical systems that interact with one another; the atmosphere, the hydrosphere, the geosphere, and the biosphere. The environmental scientists will study the interactions of all four, not just one.

We know that life on the earth depends on three interconnected factors. First is the one-way flow of energy from the sun, through living things in their feeding interactions, into the environment and eventually back into space as heat. The first and second laws of thermodynamics govern this energy flow.

Second is the cycling of matter and nutrients through parts of the biosphere. Because Earth is closed to significant inputs of matter from space, it has a fixed supply of nutrients that must be continually recycled in order to support life. The law of conservation of matter governs these biogeochemical cycles.

Third is gravity. We don't normally think of gravity as an important factor for life but it is. Gravity allows the planet to hold onto its atmosphere and helps to enable the movement and cycling of chemicals through the air, water, soil, and organisms.

The first, and most important, cycle to understand is the solar cycle. If all energy on the earth comes from the sun, how is solar energy captured and converted to a form that is useable by all organisms. Before photosynthesis evolved, organisms were mainly prokaryotic. Small, unicellular, and living in what we consider today to be harsh environments. They produced energy from the compounds found in their environment. Their ancestors are still with us today at the bottom of the oceans living in and around volcanic vents.

Photosynthesis changed all that. Now life had an unlimited source of energy and two new categories of organisms were formed: producers (autotrophs) and consumers (heterotrophs).

Producers and consumers are part of what makes up an ecosystem. They are the living (biotic) part of an ecosystem. The non-living (abiotic) part of an ecosystem is the physical characteristics of that ecosystem. Its geography, amount of water and nutrients, and temperature are all abiotic factors.

Consumers themselves are separated into a number of categories. There are herbivores (plant-eaters), carnivores (meat-eaters), omnivores (plant and animal eaters), detritivores (eat dead and decomposing matter), and decomposers.

Ecosystems also consist of food chains (who-eats-who) and food webs (a series of food chains). Both of these explain how energy flows through an ecosystem and scientists can study what happens to a food web, and ultimately the ecosystem, if food webs are disturbed by natural or man-made catastrophes. Energy pyramids also help environmental scientists plot how energy flows through an ecosystem.

There are a number of biogeochemical cycles that are important in understanding ecosystems. They are the water cycle, the carbon cycle, the nitrogen cycle, the phosphorus cycle, and the sulfur cycle. In a biology course you would study these in isolation of one another. In environmental science you need to understand how these cycles are inter-related.

Module 3 Goals

The following is a list of goals that should be met by the end of this module. The goals are very broad and, sometimes, general in nature. This was done to allow you to decide how much depth you'll want to go into for each goal.

Examining the Progress Check questions will also help focus your attention better on the goals.

1. Define ecology.
2. Describe what keeps us and other organisms alive.
3. Differentiate the major components of an ecosystem.
4. Explain what happens to energy in an ecosystem.
5. Explain what happens to matter in an ecosystem.

Module 3 Key Words

Here's a list of key words for this module:

abiotic
atmosphere
autotrophs
biogeochemical cycles
biomass
biomes
biosphere
biotic
carnivore
cell theory
cells
chemosynthesis
community
consumers
decomposers
detritivores
ecosystem
eukaryotic cell
food web
genetic diversity
geosphere
greenhouse gases
habitat
herbivore
heterotrophs
hydrosphere
limiting factor
nutrient cycles
omnivore
photosynthesis
population
producers
prokaryotic cell
species

stratosphere
trophic level
troposphere

Module 3 Progress Check

To check your understanding of the goals of this module, you should be able to answer the following questions.

1. What is a species?
2. Distinguish between a species, population, community, habitat, ecosystem, and the biosphere.
3. Distinguish between the atmosphere, troposphere, stratosphere, greenhouse gases, hydrosphere, and geosphere.
4. What three interconnected factors sustain life on earth?
5. Describe what happens to solar energy as it flows to and from the earth.
6. Define and give an example of a limiting factor.
7. Distinguish between producers (autotrophs), consumers (heterotrophs), and decomposers and give an example of each in an ecosystem.
8. Differentiate primary consumers (herbivores), secondary consumers (carnivores), high-level (third-level) consumers, omnivores, decomposers, and detritus feeders (detritivores), and give an example of each.
9. What two processes sustain ecosystems and the biosphere and how are they linked?
10. Explain what happens to energy as it flows through the food chains and food webs of an ecosystem.
11. Describe the water, carbon, nitrogen, phosphorus, and sulfur cycles and explain how human activities are affecting each cycle.

Module 4 Evolution and Biodiversity

Module 4 Overview

Biological diversity, or biodiversity, is the variety of the earth's species, the genes they contain, the ecosystems in which they live, the ecosystem processes such as energy flow, and nutrient cycling that sustains all life. Biodiversity is a vital renewable resource.

It's estimated that scientists have identified about 1.8 million of the earth's 4 million to 100 million species, and thousands of new species are identified every year. These new species include almost a million species of insects, 270,000 plant species, and 45,000 vertebrate animal species.

Species diversity is the most obvious component of biodiversity. But it's not the only component. Another important component is genetic diversity. The earth's variety of species contains a great variety of genes. Genetic diversity enables life on the earth to adapt to and survive dramatic environmental changes.

Another component is ecosystem diversity; the earth's variety of deserts, grasslands, forests, mountains, oceans, lakes, rivers, and wetlands. Each of these ecosystems are also a storehouse of genetic and species diversity.

Functional diversity is the variety of processes, such as matter cycling and energy flow, taking place within ecosystems as species interact with one another in food chains and webs.

The earth's biodiversity is a vital part of the natural capital that keeps us alive. It supplies us with food, wood, fibers, energy, and medicines. Biodiversity plays a role in preserving the quality of the air and water and maintaining the fertility of soils. It helps us to dispose of wastes and to control populations of pests. In carrying out these free ecological services, biodiversity helps to sustain life on the earth.

To understand where all this diversity, it's important to understand the principles of speciation as first explained by

Charles Darwin and Alfred Russell Wallace. You see, no organisms willingly becomes extinct. In fact, staying alive is the prevention of extinction. If a species can reproduce successfully in large enough numbers, their chances of preventing extinction remain high.

Darwin and Wallace explained the mechanisms of speciation through the principle of natural selection. The whole idea behind Darwin's work was to explain where species came from. Today, after 150 years since "On the Origin of Species" was published, we have a complete understanding of the genetic mechanisms of speciation. When we apply these mechanisms to environmental science, we see that there is a balance that all members of a population within a community and ecosystem must maintain.

Of course there are two major influences to speciation. The first is geological. We live on a dynamic planet. Volcanos and earthquakes will disrupt ecosystems. Natural disasters like forest fires and hurricanes also have the same effect. The principle of ecological succession explains how ecosystems recover, or can recover, from these catastrophes.

The other major influence is Man. Like other species, humans have survived and thrived because we have certain traits that allow us to adapt to and modify parts of the environment to increase our survival chances.

Evolutionary biologists say our success is because of three adaptations: a strong opposable thumb that allows us to grip and use tools better than, our ability to walk and run upright, and a highly evolved and complex brain.

But the adaptations that make a species successful during one period of time may not be enough to ensure the species' survival when environmental conditions change. No other species on earth has changed the environment as much as we have. Whether unintentionally or not, we have caused more damage to the earth than any other organism. Hopefully we can stop, prevent, and repair the damage before it's too late and no matter what we do it won't be enough.

Module 4 Goals

The following is a list of goals that should be met by the end of this module. The goals are very broad and, sometimes, general in nature. This was done to allow you to decide how much depth you'll want to go into for each goal.

Examining the Progress Check questions will also help focus your attention better on the goals.

1. Explain biodiversity and why it is important.
2. Describe where species come from.
3. Explain how geological processes and climate change affect evolution.
4. Clarify how speciation, extinction, and human activities affect biodiversity.
5. Describe species diversity and explain why it is important.
6. Illustrate the role species play in ecosystems.

Module 4 Key Words

Here's a list of key words for this module:

adaptation
biodiversity
biological evolution
ecological niche
endemic species
extinction
fossils
geographic isolation
keystone species
mass extinction
mutations
natural selection
reproductive isolation
speciation
species diversity

Module 4 Progress Check

To check your understanding of the goals of this module, you should be able to answer the following questions.

1. What are the four major components of biological diversity?
2. Define biological evolution and natural selection.
3. Describe the two limits to evolution by natural selection.
4. Differentiate between geographic isolation and reproductive isolation and explain how they can lead to the formation of a new species.
5. What is artificial selection and genetic engineering and give an example of each.
6. What is the difference between background extinction and mass extinction? Give examples of each.
7. What is the difference between species richness and species evenness? Give an example of each.
8. Explain why species-rich ecosystems tend to be productive and sustainable.
9. Define native, nonnative, indicator, keystone, and foundation species and give an example of each.
10. Explain why birds are excellent indicator species.

Module 5 Species Interactions and Populations

Module 5 Overview

Ecologists have identified five types of interactions between species that share limited resources such as food, shelter, and space.

Interspecific competition occurs when members of two or more species interact to gain access to the same limited resources such as food, light, or space.

In predation, the predator feeds directly on all or part of the prey. Together, the two different species, such as wolves (the predator or hunter) and rabbits (the prey or hunted), form a predator–prey relationship.

Parasitism occurs the parasite feeds on the body of, or the energy used by, the host, usually by living on or in the host. In this relationship, the parasite benefits and the host is harmed but not immediately killed. Unlike the typical predator, a parasite usually is much smaller than its host and rarely kills its host. Also, most parasites remain closely associated with their hosts, draw nourishment from them, and may gradually weaken them over time. Some parasites, such as tapeworms and some disease-causing microorganisms, live inside their hosts. Other parasites attach themselves to the outsides of their hosts.

Mutualism is an interaction that benefits both species by providing each with food, shelter, or some other resource.

Commensalism is an interaction that benefits one species but has little, if any, effect on the other.

The most common interaction between species is competition for limited resources. While fighting for resources does occur, most competition involves the ability of one species to become more efficient than another species in acquiring food or other resources.

Humans, on the other hand, compete with many species for space, food, and other resources. As our ecological footprints grow and spread and we convert more of the earth's land, aquatic resources, and productivity to our uses, we are taking over, or destroying, the habitats of other species and denying them the resources they need to survive.

All populations have certain characteristics. Populations differ in factors such as their distribution, numbers, age structure, and density. The study of how the characteristics of populations change in response to changes in environmental conditions is called population dynamics. These conditions might include temperature, presence of disease organisms or harmful chemicals, resource availability, and arrival or disappearance of competing species.

Individuals of a species in many populations live together in clumps or patches. There are several reasons for this. First, resources will vary greatly in availability from place to place, so the species will cluster where resources are available. Second, organisms moving in groups have a better chance of coming across patches of resources, such as water and vegetation, than they would searching for the resources on their own. Third, living in groups protects the population better from predators. Fourth, hunting in packs gives some predators a better chance of finding and catching prey. Fifth, some species form temporary groups for mating and caring for young.

There are four variables that govern the size of a population: births, deaths, immigration, and emigration. If the number of births is greater than the number of deaths, a population will increase in size. A population will decrease if the number of deaths is greater than the number of births. Immigration is the number of individuals that move into the population, emigration is the number of individuals that leave a population. So the growth rate of any population is equal to births and immigration subtracted from death and emigration.

A population's age structure is the proportion of individuals at various ages. Age structure diagrams describe organisms not mature enough to reproduce, those capable of reproduction, and those too old to reproduce.

Any population of organisms cannot grow indefinitely. There are many factors that influence over-all growth rates of populations. Core to the growth rate is availability of resources. There are limiting factors that control the rate of growth. Density-dependent factors would include light, water, space, nutrients, or exposure to too many competitors, predators, or infectious diseases. The growth rate will depend on the size of the population.

There are also factors that are independent of population density. Density-independent factors would include factors such as weather (hurricanes, tornados, floods, drought, etc), natural disasters (volcanos, earthquakes, forest fires), and human influences (destruction of habitats, pollution, and human population growth).

Population growth can be illustrated graphically by a graph that shows exponential growth, which is rapid, uncontrollable growth up to a point where the available resources can no longer support the population, the carrying capacity (K). This is followed by logistic growth, which is a decrease in the population until the population levels off. These graphs are referred to as the J-curve (exponential growth) and the S-curve (cyclical increase and decrease).

Small, isolated populations can be affected by the lack of genetic diversity in a number of ways. One such factor is called the founder effect and occurs when a few individuals in a population colonize a new habitat that is geographically isolated from other members of the population.

Another factor is called a demographic bottleneck. It occurs when a few individuals in a population survive a catastrophe such as a fire or hurricane. Lack of genetic diversity, in this case, limits the ability of the individuals to rebuild the population and may lead to an increase in the frequency of harmful genetic diseases.

A third factor is genetic drift. This involves random changes in the gene frequencies in a population that can lead to unequal reproductive success. Some individuals, for example, may breed more frequently than others and their genes may eventually dominate the gene pool of the population. The founder effect is one cause of genetic drift.

A fourth factor is inbreeding. It occurs when individuals in a small population mate with one another. This can occur through a demographic bottleneck and can also increase the frequency of defective genes.

Ecosystems respond to changes in environmental conditions through ecological succession. There are two types of succession: primary and secondary. Primary succession begins with an essentially lifeless area where there is no soil in a terrestrial system or bottom sediment in an aquatic system. Primary succession usually takes a long time because there is no fertile soil to provide the nutrients needed to establish a plant community.

The slow process of soil formation begins when pioneer species arrive and attach themselves to inhospitable patches of the weathered rock. These early plant species start the process of soil formation by trapping wind-blown soil particles and tiny pieces of detritus, and adding their own wastes and dead bodies. They also secrete mild acids that further fragment and break down the rock.

After thousands of years, the soil may be deep and fertile enough to store moisture and nutrients to support the growth of mid-successional plant species. As these tree species grow and create shade, they are replaced by late successional plant species that can tolerate shade. Unless fire, flooding, severe erosion, tree cutting, climate change, or other natural or human processes disturb the area, what was once bare rock becomes a complex forest community or ecosystem.

Secondary succession occurs in an area where an ecosystem has been disturbed, removed, or destroyed, but some soil remains. Examples of secondary succession include abandoned farmland, burned or cut forests, heavily polluted streams, and land that has been flooded. In the soil that remains on disturbed land systems, new vegetation can germinate, usually within a few weeks, from seeds already in the soil and from those imported by wind, birds, and other animals.

It's important to note that there are limits to the stresses that any ecosystem can take, whether is be a local ecosystem or a global one. All ecosystems can reach a tipping point. This occurs when any additional stress can cause the system to change in an abrupt and irreversible way that often involves the collapse of the ecosystem. As an example, environmental scientists agree that continual release of carbon dioxide into the atmosphere and the oceans will lead to a tipping point where no matter what we do to decrease the amount of carbon dioxide, the damage has been done and is irreversible.

Module 5 Goals

The following is a list of goals that should be met by the end of this module. The goals are very broad and, sometimes, general in nature. This was done to allow you to decide how much depth you'll want to go into for each goal.

Examining the Progress Check questions will also help focus your attention better on the goals.

1. Explain how species interact.
2. Describe how natural selection can reduce competition between species.
3. Describe the limits to the growth of populations.
4. Explain how communities and ecosystems respond to changing environmental conditions.

Module 5 Key Words

Here's a list of key words for this module:

age structure
carrying capacity
coevolution
commensalism
competition
ecological succession
exponential growth
logistic growth
mutualism
parasitism
population dynamics
predation
predator-prey relationship
primary succession
resource partitioning
secondary succession

Module 5 Progress Check

To check your understanding of the goals of this module, you should be able to answer the following questions.

1. Define interspecific competition, predation, parasitism, mutualism, and commensalism and give an example of each.
2. How do each of these species interactions affect the population sizes of species in ecosystems?
3. What is a predator–prey relationship and give an example.
4. Describe four ways in which prey species can avoid their predators and four ways in which predators can capture these prey.
5. Define and give an example of coevolution.
6. What is population dynamics and explain why most populations live in clumps?
7. What are the four variables that govern changes in population size and write an equation showing how they interact.
8. What is a population's age structure and what are three major age group categories?
9. Define population density and explain how it can affect the size of some but not all populations.
10. Explain why humans are not exempt from nature's population controls.
11. Distinguish between primary ecological succession and secondary ecological succession and give an example of each.
12. Explain how living systems achieve some degree of stability or sustainability by undergoing constant change in response to changing environmental conditions.

Module 6 Human Population Study

Module 6 Overview

The subject of human population growth has been a topic of discussion not only in scientific circles, but in political circles for decades. Paul Ehrlich predicted the end of human civilization by the end of the 1980s if population growth wasn't checked in his wildly popular and acclaimed book, "The Population Bomb." Published in 1971, Ehrlich claimed that once global population hit 3 billion, there would be a total collapse of human population. Since we're at over 7 billion today, he may have missed his mark somewhat.

Buy how many people can the earth support? For most of human history, population grew slowly. For the past 200 years, though, the population has increased rapid growth reflected in the characteristic J-curve.

There are three factors for this increase. First, European explorers found the New World and they emigrated to these new and fertile lands. Second, the development of modern agriculture allowed more people to be fed. With the richness of the land in the new world, those settlers could farm on soil that was much more productive than what they left behind in Europe. And the farms in the New World were self-sufficient. Third, the history of the Middle Ages and the Renaissance is marked with epidemic after epidemic. And as communities and cultures continued to become more and more diverse, the number and power of the epidemics rose. It's estimated that the population of Europe decreased by as much as 50% as a result of the Plague. So the development of sanitation systems, antibiotics, and vaccines helped control infectious disease agents. As a result, death rates dropped sharply below birth rates and population size grew rapidly.

About 10,000 years ago there were about 5 million humans on the planet; now there are over 7 billion of us. It took from the time we arrived until about 1927 to add the first 2 billion people to the planet; less than 50 years to add the next 2 billion (1974); and just 25 years to add the next 2 billion (1999). Now we are trying to support 7 billion people and perhaps as many as 10-12 billion by 2050.

The question is, what is our carrying capacity? We have artificially manipulated our environment to develop better ways to feed more people. We have developed medicines and medical practices to keep people alive longer. And we continually try to educate people about the cost of large families.

So the good news is that the rate of population growth has slowed. But maybe we passed the tipping point. Because the world's population is still growing exponentially at a rate of 1.22% a year. This means that an average of nearly 225,000 people each day, or 2.4 more people every time your heart beats.

Geographically, this growth is unevenly distributed. The world's developed countries are growing at a rate of 0.1% a year. The developing countries are growing 15 times faster with a rate of 1.5% a year. Most of the world's population growth takes place in already heavily populated parts of world, most of which are the least equipped to deal with the pressures of such rapid growth.

How many of us will there be in 2050? Current estimates puts that number in the range of 9-12 billion people. This depends on projections about the average number of babies women are likely to have. According to many, this is the carrying capacity and the scenarios of what happens after that point run from scary to really scary.

Two excellent sources of information on human population growth are the population Reference Bureau (www.prb.org) and the CIA World Factbook (http://1.usa.gov/194Uu5A). PBS also produced a program called "World in the Balance" which is also excellent at illustrating the problem of overpopulation.

Finding a solution isn't easy. Scientists can explain the biology behind population dynamics and show all the J-curves they want. But unfortunately there are many cultural, religious, economic, and political forces at play. What we do know is that developed countries who are stable politically, economically, and socially are growing at much slower rates than the rest of the world. There are countries in Western Europe that are showing negative growth rates. More people are dying than are born.

At the other end of the spectrum are the underdeveloped countries of the world. Niger, for example, currently has a population of over 16 million. When you look at the age-structure diagram of Niger, you see where this country is going in 10-15 years. Of the 16 million people in Niger, 50% are under the age of 14. When all of those children are 14 or older, Niger will have serious economic, political, and social problems on there hands.

Even India, who is a developing country, is facing serious problems. India currently has over 1.2 billion people with 30% of its population under the age of 14. India, the world's largest democracy, has struggled with is exploding population for many decades with seemingly little effect.

China, on the other hand, has done a great deal of work in trying to curb its population growth, even though it generates great controversy. China has a population of 1.3 billion people with only 17% under the age of 14. China's policies have been effective in curbing its population, but at a human cost.

Over the years Paul Ehrlich has been marginalized by the scientific community. Some claim he became too outrageous in his belief of the population bomb. Maybe he was 80 years too early.

Module 6 Goals

The following is a list of goals that should be met by the end of this module. The goals are very broad and, sometimes, general in nature. This was done to allow you to decide how much depth you'll want to go into for each goal.

Examining the Progress Check questions will also help focus your attention better on the goals.

1. Describe how many people the earth can support.
2. Define the factors that influence the size of the human population.
3. Explain how a population's age structure affects its growth or decline.
4. Explain ways we can slow human population growth.

Module 6 Key Words

Here's a list of key words for this module:

birth rate
cultural carrying capacity
death rate
demographic transition
fertility rate
infant mortality rate
life expectancy
migration
population change

Module 6 Progress Check

To check your understanding of the goals of this module, you should be able to answer the following questions.

1. List at least three factors that account for the rapid growth of the world's human population over the past 200 years.
2. What is the cultural carrying capacity of a population?
3. Differentiate between crude birth rate and crude death rate.
4. Define fertility rate.
5. What is the difference between replacement-level fertility rate and total fertility rate?
6. Distinguish between life expectancy and infant mortality rate and explain how they affect the population size of a country.
7. What is the difference between immigration and emigration?
8. What is the age structure of a population and explain how it affects population growth and economic growth.
9. What is the demographic transition and what are its four stages?
10. How has human population growth interfered with natural processes related to three of the scientific principles of sustainability?

Unit 3 - Biodiversity and Sustainability

Module 7 Biomes and Climate

Module 7 Overview

Earth is unique among the planets of the solar system. We have weather. Because we have weather, we have climate and climate is the driving force behind the formation of biomes.

First, you need to differentiate between weather and climate. Weather is short-term and occurs in a local area and is measured over hours or days. Climate is an area's general pattern of weather conditions measured over long periods of time ranging from decades to thousands of years.

Climate varies in different parts of the world because of four factors. First is the uneven heating of the earth's surface by the sun. Air is heated much more at the equator, where the sun's rays strike directly, than at the poles, where sunlight strikes at a slanted angle and spreads out over a much greater area. These differences help explain why tropical regions near the equator are hot, why polar regions are cold, and why temperate regions in between generally have intermediate average temperatures.

Second are the properties of air, water, and land. Solar energy evaporates ocean water and transfers heat from the oceans to the atmosphere, especially near the equator. The evaporation of this water creates giant cyclical convection cells that circulate air, heat, and moisture both vertically and from place to place in the atmosphere. Remember, hot air rises because it is less dense and cold air sinks because it is more dense.

Third is the rotation of the earth on its axis. As the earth rotates around its axis, the equator is spinning faster than the poles. As a result, heated air currents rising above the equator and moving north and south to cooler areas are deflected to the west or east. The atmosphere over these different areas is divided into huge regions called cells and the differing directions of air movement are called prevailing winds.

The fourth is the angle of the earth's axis and the shape of its orbit around the Sun. The earth's axis is tilted 23.5 degrees. This angle results in further heating and cooling of the air in regions on the planet. It also provides for the 6 months of light and darkness at each of the poles. The angle is the prime reason why we have seasons. The elliptical course of the earth's path around the Sun also affects climate. Winters and summers in the southern hemisphere tend to be more extreme than in the northern hemisphere since the earth is farthest or closest to the Sun respectively.

In addition to all the above, you have to consider the surface of the planet. Oceans cover over 75% of the surface of earth. Water heats slowly and retains heat longer than land. Mountains will block the flow of prevailing surface winds causing one side of mountain ranges to be moist. Think of the lushness around Fresno and Stockton in California on the western side of the Sierra Nevada mountains. And because the Sierra Nevada block the westerly winds off the Pacific Ocean the western side of the Sierra Nevada are dry, Death Valley dry.

All of these factors come into play in the wide variety of biomes on earth. A biome is a large group of communities of organisms that have adapted to a specific habitat. We have two categories of biomes: terrestrial and aquatic.

The following are brief descriptions of the major world biomes.

Desert Biomes

What defines a desert is a combination of low rainfall and different average temperatures. This creates tropical, temperate, and cold deserts.

Tropical deserts are hot and dry most of the year. They have few plants and a hard, windblown surface with scattered rocks and sand. They are the deserts we often see in the movies.

In temperate deserts, daytime temperatures are high in summer and low in winter and there is more precipitation than in tropical deserts. There is little vegetation that consists mostly of widely dispersed, drought-resistant shrubs and other succulents adapted to the lack of water and temperature variations. The Mojave Desert in California is an example of this type of desert.

In cold deserts vegetation is sparse. Winters are cold, summers are warm or hot, and precipitation is low. Desert plants and animals have adaptations that help them to stay cool and to get enough water to survive. The Gobi desert in Mongolia is a cold desert.

All desert ecosystems are fragile. The soils can take decades to recover from disturbances. This is because of slow plant growth, low species diversity, slow nutrient cycling, and lack of water.

Grassland Biomes

Grasslands occur mostly in the interiors of continents in areas that are too moist for deserts and too dry for forests. The three main types of grassland result from combinations of low average precipitation and various average temperatures.

Tropical savannas will have herds of grazing and browsing hoofed animals, and their predators. Herds of the grazing and browsing animals will migrate to find water and food in response to seasonal and year-to-year variations in rainfall and food availability.

A savanna is a type of tropical grassland that contains widely scattered clumps of trees such as acacia, which are covered with thorns that help to keep herbivores away. This biome usually has warm temperatures year-round and alternating dry and wet seasons.

Temperate grassland have winters that are bitterly cold, summers that are hot and dry, with little annual precipitation that falls unevenly through the year. Because the aboveground parts of most of the grasses die and decompose each year, organic matter accumulates to produce a deep, fertile soil. This soil is held in place by a thick network of intertwined roots of drought-tolerant grasses. The natural grasses are also adapted to fires, which burn the plant parts above the ground but do not harm the roots, from which new grass can grow.

There are two types of temperate grasslands: tall-grass prairies and short-grass prairies. These can be found in Midwestern and western United States and Canada. Short-grass prairies get about 10 inches of rain a year, and tall-grass prairies can get up 35 inches.

Cold grasslands, usually called arctic tundra, are found south of the arctic polar ice cap. During most of the year, these treeless plains are bitterly cold, swept by frigid winds, and covered by ice and snow. Winters are long and dark, and what little precipitation there is falls as snow.

This biome is carpeted with a thick, spongy mat of low-growing plants, primarily grasses, mosses, lichens, and dwarf shrubs. Trees and tall plants cannot survive in the cold and windy tundra because they would lose too much of their heat. Animals in cold grasslands survive the intense winter cold through adaptations such as thick coats of fur or feathers, or living underground.

Temperate Shrublands

These biomes (also called chaparral) are found in many coastal regions that border on deserts. Closeness to the ocean provides a slightly longer winter rainy season than nearby temperate deserts, and fogs during the spring and fall that reduce evaporation. These biomes can be found along coastal areas of southern California, the Mediterranean Sea, central Chile, southern Australia, and southwestern South Africa.

Chaparral consists mostly of dense growths of low-growing evergreen shrubs and occasional small trees with leathery leaves that reduce evaporation. The soil is thin and not very fertile. Animal species of the chaparral include mule deer, chipmunks, jackrabbits, lizards, and a variety of birds.

Forests

Forest biomes are dominated by trees and consist of three main types: tropical, temperate, and cold (northern coniferous and boreal). Forests result from combinations of the precipitation level and various average temperatures.

Tropical rain forests are found near the equator, where hot, moisture-laden air rises and dumps its moisture. These lush forests have year-round, uniformly warm temperatures, high humidity, and heavy rainfall almost daily. This fairly constant warm and wet climate is ideal for a wide variety of plants and animals.

Temperate deciduous forests are found in areas with moderate average temperatures that change significantly with the season. These areas have long, warm summers, cold but not too severe winters, and abundant precipitation, often spread fairly evenly throughout the year. This biome is dominated by a few species of broadleaf deciduous trees such as oak, hickory, maple, poplar, and beech. They survive cold winters by dropping their leaves in the fall and becoming dormant through the winter.

Evergreen coniferous forests are also called boreal forests and taigas. These cold forests are found just south of the arctic tundra in northern regions across North America, Asia, and Europe and above certain altitudes in the High Sierra and Rocky Mountains of the United States. In this subarctic climate, winters are long, dry, and extremely cold. Summers are short, with cool to warm temperatures.

Most boreal forests are dominated by a few species of evergreen trees such as spruce, fir, cedar, hemlock, and pine that keep most of their narrow-pointed needles year-round. The small, needle-shaped, waxy-coated leaves of these trees can withstand the intense cold and drought of winter. These trees can take advantage of the brief summers in these areas without taking time to grow new needles. Plant diversity is low because few species can survive the winters when soil moisture is frozen.

Mountains

Though not a biome, the importance of mountains cannot be overlooked because they play important ecological roles. Mountains contain the majority of the world's forests, which are habitats for much of the planet's terrestrial biodiversity. They often provide habitats for species found nowhere else on earth. They also serve as sanctuaries for animal species driven to migrate from lowland areas to higher altitudes.

Mountains help to regulate the earth's climate. Mountaintops covered with ice and snow affect climate by reflecting solar radiation back into space. This helps to cool the earth and offset global warming. Many of the world's mountain glaciers are melting, it is believed, mostly because of global warming. While glaciers reflect solar energy, the darker rocks exposed by melting glaciers absorb that energy. This helps to increase global warming, which melts more glaciers and warms the atmosphere more.

Mountains can affect sea levels by storing and releasing water in glacial ice. As the earth gets warmer, mountaintop glaciers and other land-based glaciers can melt, adding water to the oceans and helping to raise sea levels.

Lastly, mountains play an important role in the water cycle by serving as a major storehouse of water. In the warmer weather of spring and summer, much of their snow and ice melts and is released to streams for use by wildlife and by humans for drinking and irrigating crops.

Module 7 Goals

The following is a list of goals that should be met by the end of this module. The goals are very broad and, sometimes, general in nature. This was done to allow you to decide how much depth you'll want to go into for each goal.

Examining the Progress Check questions will also help focus your attention better on the goals.

1. Describe the factors that influence climate.
2. Explain how climate affects the nature and locations of biomes.
3. Describe how humans have affected the world's terrestrial ecosystems.

Module 7 Key Words

Here's a list of key words for this module:

climate
currents
desert
forest systems
grasslands
greenhouse effect
permafrost
rain shadow effect
weather

Module 7 Progress Check

To check your understanding of the goals of this module, you should be able to answer the following questions.

1. What is the difference between weather and climate?
2. Define the three major factors that determine how air circulates in the lower atmosphere.
3. Describe how the properties of air, water, and land affect global air circulation.
4. What is the greenhouse effect and why is it important to the earth's life and climate?
5. What is a biome?
6. Explain why there are three major types of each of the major biomes (deserts, grasslands, and forests).
7. How does climate and vegetation vary with latitude and elevation?
8. Why have many of the world's temperate grasslands disappeared?
9. Why is biodiversity so high in tropical rain forests?
10. What are the connections between the earth's winds, climates, and biomes and the four scientific principles of sustainability?

Module 8 Extinction

Module 8 Overview

Extinction lies at the very core of life. The sole function of every species is to prevent extinction. No species intentionally become extinct. They do so because of many reasons not always in their control.

We are well aware of extinction today because of the attention the media gives it. Anyone who has paid attention to environmental issues over the years knows that there are many species today who are on the verge of extinction. Many of these extinctions are, unfortunately, because of human activities. But this needs to be put in perspective. Ninety-nine percent of all the species that have ever lived on earth are extinct. What we see today represents only 1% of all the life that has ever lived on earth. That's a dramatic illustration of how evolution, through natural selection, works! Adapt or die.

As much as environmentalists will try to convince us that human activity is responsible for the extinction of many species, and we are, the numbers are insignificant when compared to all the others that have gone before us. The single most successful group of organisms that has ever lived on earth are, of course, the dinosaurs. They dominated all other organisms for over 150 million years. And they became extinct, for lots of reasons including the earth getting hit by a massive asteroid. I don't know about you, but I'm not taking the blame for the dinosaurs since humans were about 63 millions years in the dinosaur's future.

Environmental scientists have calculated what is known as the background extinction rate. In other words, the normal rate of extinction before human interference. That rate has been calculated at 0.0001%, or 1 species for every million.

We also know, and this is supported by the fossil record, that there have been at least five mass extinctions. Periods in geologic time where most of the life on the planet became extinct. The last one, of course, occurred 65 million years ago with the extinction of not only the dinosaurs, but with about 80% of all other life. Asteroids have a tendency to do that.

Scientists distinguish between local, ecological, and biological extinctions. In local extinctions, species can no longer be found in an area it once inhabited, but can be found in other habitats. Ecological extinction means that there are so few members of a species in a habitat that they can no longer play their ecological roles in the habitats they are living in. The most serious level is biological extinction. When we hear that a species is extinct, this is what they are referring to. The species can no longer be found anywhere on earth. They are gone, along with their genetic uniqueness.

There is no question that human activity has altered many ecosystems, that we have destroyed many habitats, and have hunted animals into extinction. The current background extinction rate is about 0.01%. That's a dramatic increase from the original rate.

If all species eventually become extinct, what's all the concern about premature extinctions? Does it really matter that the passenger pigeon became prematurely extinct because of human activities, or that the remaining orangutans or some unknown plant or insect in a tropical forest might suffer the same fate? Why should we care if we speed up the extinction rate over the next 50–100 years? Quite simply because it takes 5–10 million years for natural speciation to rebuild the biodiversity that has been destroyed.

The greatest threats to any species are, in order, loss or degradation of its habitat, harmful invasive species, human population growth, pollution, climate change, and overexploitation. According to researchers, the greatest threat to any wild species is habitat loss, degradation, and fragmentation.

Deforestation in tropical areas is the greatest eliminator of species, followed by the destruction and degradation of coral reefs and wetlands, plowing of grasslands, and pollution of streams, lakes, and oceans. Temperate biomes have been affected more by habitat loss and degradation than have tropical biomes because of widespread economic development in these parts of the world.

Habitat fragmentation (by roads, logging, agriculture, and urban development) occurs when a large area of habitat is reduced in area and divided into smaller, more scattered, and isolated patches, or habitat islands. This process decreases tree populations in forests, blocks migration routes, and divides populations of a species into smaller and more isolated groups that are more vulnerable to predators, competitor species, disease, and catastrophic events such as storms and fires.

The best way to prevent the premature extinction of organisms is to enact laws that are designed to prevent it from happening. Of course this works well in developed countries. But what happens in those countries that have emerging economies, or in those underdeveloped countries that are economically and politically unstable?

In 1973, the United States enacted the Endangered Species Act. It is the most far-reaching piece of legislation to protect wild-life ever enacted by any government. In 1973, there were 92 endangered species protected by the law. By 2007 that number increased to 1,350.

Critics of the law say that it should be weakened or even repealed. They claim that the rights of humans have to be held higher that the rights of plants and animals. This point raises the question of does a government have the moral and social imperative to protect it's natural resources and if they do, how far to they go in this protection?

Some argue that, because we have identified fewer than 2 million of the estimated 5–100 million species on the earth, it makes little sense to take drastic measures to preserve them. But scientists disagree with this logic. They remind us that the earth's species are the primary components of its biodiversity. This biodiversity should not be degraded because of the economic and ecological services it provides. Scientists urge us to use great caution in making potentially harmful changes to ecosystems and to take precautionary action to help prevent serious environmental problems, including premature extinctions.

Module 8 Goals

The following is a list of goals that should be met by the end of this module. The goals are very broad and, sometimes, general in nature. This was done to allow you to decide how much depth you'll want to go into for each goal.

Examining the Progress Check questions will also help focus your attention better on the goals.

1. Explain the role humans play in the premature extinction of species.
2. Describe why we should care about preventing premature species extinction.
3. Identify how humans accelerate species extinction.
4. Explain how we protect wild species from extinction resulting from our activities.

Module 8 Key Words

Here's a list of key words for this module:

biodiversity hotspots
deforestation
ecological restoration
old-growth forest
overgrazing
pastures
rangelands
second-growth forest
tree farm

Module 8 Progress Check

To check your understanding of the goals of this module, you should be able to answer the following questions.

1. What is the difference between background extinction and mass extinction?
2. Define the extinction rate of a species.
3. Differentiate between endangered species and threatened species.
4. Many extinction experts believe that human activities are now causing a sixth mass extinction. State four reasons why they believe this.
5. What are the six largest causes of premature extinction of species resulting from human activities?
6. What is habitat fragmentation, and how does it threaten many species?
7. What are two examples of the harmful effects of nonnative species that have been introduced (a) deliberately and (b) accidentally?
8. List ways to limit the harmful impacts of nonnative species.
9. Describe two international treaties that are used to help protect species.
10. Describe the U.S. Endangered Species Act, how successful it has been, and the controversy over this act.

Module 9 Maintaining Land Biomes

Module 9 Overview

About 30% of the earth is covered by forests. Of this about 35% are old-growth forests; those forests that has not been touched by human activity for over 200 years. Most of these forests occur in Russia, Canada, and Brazil. Another 60% of the above are new-growth forests; those forests that are the result of secondary succession. These forests develop as a result of natural forces like fires or volcanos and human activity. The remaining 5% are tree plantations. As the name indicates, these forests are commercially grown primarily for the lumber or pulp. The United States has more tree plantations than any other country.

We cannot overstate the importance of the ecological role that tress play. Ever wonder where the oxygen comes from that we breathe? Go out and thank a tree. Do you ever wonder what happens to all the carbon dioxide that we breathe out goes? Go out and than that same tree. So in addition to providing us with the oxygen we breathe, trees remove carbon dioxide from the air and that helps stabilize the earth's temperature. In an age where we are concerned about climate change on a large scale, we should all be given shovels and told to plant trees.

Aside from fact that trees are important for living, they play other roles equally as important. For example, traditional medicines, used by 80% of the world, come from natural plants in forests. The chemicals found in tropical forest plants are used as blueprints for making most of the world's prescription drugs. Forests are also habitats for about two-thirds of the earth's terrestrial species. Lastly, one of every four people depend on forests for their livelihoods.

The single largest threat to forests is uncontrollable logging. Throughout history, people understood the importance of controlling the destruction of their forests. There are many references to societies collapsing because of uncontrollable logging. And the very first environmental laws enacted were to protect the trees. Wood was used for everything in those ancient times, and not in the too distant past either, and since trees seem to be everywhere it must have seemed like an inexhaustible supply.

And not much has changed over the centuries. You would think that with all the emphasis placed on protecting this vital natural resource that governments would do their best to protect the forests. But this is not the case. More than 70 countries around the world participate in uncontrollable logging. In Kenya, for example, 37 of the 41 national parks have been devastated by uncontrollable logging. And China, which has destroyed nearly all of its own forests, imports more wood than any other country. Most of this wood is cut illegally and is used for many products produced in China.

Unsustainable and illegal logging impacts the ecosystems tremendously. From the roads needed to access the forest to the destruction of the local habitats and loss of animal life found in those habitats, to the eventual erosion of the soil. What is left behind is land that is of little use to anyone.

Amazingly, with all this knowledge of how important preserving the forests is, the earth loses around 50,000 square miles of forests every year. That's about the size of the state of Mississippi. Most of this destruction occurs in Latin America, Africa, and Indonesia, where most of the tropical rain forests are located.

With the continual destruction of the world's forests, at some point we are going to reach a tipping point, a point of no return. If we continue to lose 50,000 square miles of forests each year (which will more than likely increase) and with continual increase in the human population, at what point is the amount of oxygen produced going to be insufficient to support the world's population? Not that we'll all suffocate immediately, but we'll begin to witness the slow decrease in the level of atmospheric oxygen, which is 21%. Add to that the increased levels of carbon dioxide since we have fewer trees to filter it out and more people who breathe it out. Countries around the world are united in their efforts to stop the illegal destruction of the forests and to protect and manage the remaining forests from further degradation.

The federal government owns 29% of all the land in the United States. These 655 million acres are managed not only by the Park Service, but also the Forest Service, the Bureau of Land Management, and the Fish and Wildlife Service. The U. S. Park

Service was established in 1912 to manage the national parks. President Teddy Roosevelt was the first president to establish the idea of federally protected lands in order to preserve our natural resources.

Ecologically, the national parks are under great stress from human activities and pollution. You would think that these federally protected lands should be in pristine condition but they are not. People visit the national parks by the millions every year. They bring their cars, SUV's and campers. The leave behind garbage, feed the bears and chipmunks, and they sometimes start fires. And they pump millions of dollars into the local economies. It's estimated that it will take over 6 billion dollars to repair the national parks. That, of course doesn't include the yearly operating costs. It will take desire, creative thinking, and innovation to be sure that the parks do not continue to decline. This is probably not what Teddy Roosevelt had in mind when he established the first national park.

The other issue with federally owned lands is the sale of these lands to private companies without any stipulation to their condition. For example, under the U.S. Mining Act of 1872, mining companies can purchase federal land cheaply, extract millions of dollars worth of minerals from the soil, abandon it when their done and not be responsible for the environmental mess they leave behind.

The intentions of the government to protect the land may be good, the influence that is lobbied by self-serving parties has ended up in creating more problems. The solutions have to come from the people. Time and time again we see the success of individual environmental groups who raise enough awareness to a call for action that change happens. The only way the land is going to be protected is when enough people organize and force change from the bottom up.

Module 9 Goals

The following is a list of goals that should be met by the end of this module. The goals are very broad and, sometimes, general in nature. This was done to allow you to decide how much depth you'll want to go into for each goal.

Examining the Progress Check questions will also help focus your attention better on the goals.

1. Describe the major threats to forest ecosystems.
2. Explain how we should manage and sustain forests.
3. Explain how we should manage and sustain parks and nature reserves.
4. Describe what the ecosystem approach to sustaining biodiversity is.

Module 9 Key Words

Here's a list of key words for this module:

background extinction
endangered species
extinction rate
instrumental value

Module 9 Progress Check

To check your understanding of the goals of this module, you should be able to answer the following questions.

1. Distinguish between an old-growth forest, a second-growth forest, and a tree plantation (tree farm or commercial forest).
2. What are the major ecological and economic benefits that forests provide?
3. Differentiate between selective-cutting, clear-cutting, and strip-cutting in the harvesting of trees.
4. What are some ecological benefits of occasional surface fires?
5. Define deforestation and list some of its major harmful environmental effects.
6. What are the major basic and secondary causes of tropical deforestation?
7. What are three ways to reduce the need to harvest trees?
8. State five ways to protect tropical forests and use them more sustainable.
9. What major environmental threats affect national parks?
10. What is a biological hotspot and why is it important to protect such areas?

Module 10 Aquatic Biodiversity

Module 10 Overview

One of the most famous pictures ever taken of earth is called "Earthrise". Taken by the astronauts of Apollo 8 as they came around the back side of the Moon for the first time, it shows how much of the planet is water. Suspended in the blackness of space is this little blue marble.

About 71% of the earth's surface is covered by saltwater and another 2% by freshwater. In proportion to the size of the entire planet, it all amounts to a thin and precious film of water.

The global oceans are a single and continuous body of water. For the purposes of geography it is divide into four large areas; the Atlantic, Pacific, Arctic, and Indian Oceans. The largest ocean is the Pacific, which contains more than half of the earth's water and covers one-third of the earth's surface.

Instead of referring to them as biomes, we refer to the oceans as aquatic life zones. The distribution of aquatic life is determined largely by the water's salinity; the amounts of various salts dissolved in a given volume of water. Aquatic life zones are therefore classified into two major types: marine and freshwater (lakes, rivers, streams, and inland wetlands). Some systems such as estuaries are a mix of saltwater and freshwater are classified as marine systems. Aquatic systems play vital roles in the earth's biological productivity, climate, biogeochemical cycles, and biodiversity. They also provide us with fish, shellfish, minerals, recreation, and transportation routes.

Both saltwater and freshwater life zones contain several major types of organisms. One type consists of weakly swimming, free-floating plankton. A second major type of organisms is called nekton, strongly swimming consumers such as fish, turtles, and whales. The third type, called benthos, consists of bottom dwellers such as oysters, clams and worms, and lobsters and crabs. A fourth major type are the decomposers, which break down organic compounds in the dead bodies and wastes of aquatic organisms into nutrients that can be used by aquatic primary producers.

Most forms of aquatic life can be found in either the surface, middle, or bottom layers of salt- and freshwater systems. The key factors in determining the types and numbers of organisms found in these layers are temperature, dissolved oxygen, availability of food, and availability of light and nutrients required for photosynthesis.

In deep aquatic systems, photosynthesis is largely confined to the upper layer through which sunlight can penetrate. This layer is referred to as the euphotic or photic zone. The depth of the euphotic zone in oceans and deep lakes can be reduced when the water is clouded by excessive algal growth. This cloudiness is called turbidity and occurs naturally, or can result from disturbances like the clearing of land, which causes silt to flow into bodies of water. One of the problems coral reefs are having today is excessive turbidity due to silt runoff.

Marine Systems

Marine aquatic systems are enormous reservoirs of biodiversity. The many different ecosystems host a great variety of species, genes, and biological and chemical processes. Marine life can be found in one of three major zones: the coastal zone, the open sea, and the ocean bottom.

Coastal Zone

The coastal zone is the shallow water that extends from the high-tide mark on land to the gently sloping, shallow edge of the continental shelf. This zone contains water that is warm and nutrient-rich. The coastal zone makes up less than 10% of the world's oceans, but contains 90% of all marine species and is the site of most large commercial marine fisheries. Coastal zone aquatic systems consists of estuaries, coastal wetlands, mangrove forests, and coral reefs.

Estuaries are found where rivers meet the sea. These are partially enclosed bodies of water where seawater mixes with freshwater along with nutrients and pollutants from streams and rivers.

Estuaries, along with coastal wetlands (coastal land areas covered with water all or part of the year) include river mouths, inlets, bays, sounds, and salt marshes in temperate zones, and mangrove forests in tropical zones. Because of high nutrient inputs from rivers and nearby land, rapid circulation of nutrients by tidal flows, and ample sunlight penetrating the shallow waters, these are some of the earth's most productive ecosystems.

The tropical equivalent of salt marshes are mangrove forests. They can be found along 70% of gently sloping sandy coastlines in tropical and subtropical regions. The dominant organisms in these nutrient-rich coastal forests are, of course, mangroves. These plants represent 69 different species of trees that can grow in salt water. They have extensive root systems that often extend above the water, where they can obtain oxygen and support the trees during periods of changing water levels.

These aquatic systems provide important ecological and economic services. First, they help to maintain water quality in tropical coastal zones by filtering pollutants, excess plant nutrients, and sediments, and by absorbing other pollutants. Secondly, they provide food, habitats, and nursery sites for a variety of aquatic and terrestrial species. Third, they reduce storm damage and coastal erosion by absorbing waves and storing excess water produced by storms and tsunamis. Finally, they have historically supplied timber and fuelwood to coastal communities.

Coral reefs are the world's oldest, most diverse, and most productive ecosystems. They also provide homes for one-fourth of all marine species. These centers of aquatic biodiversity are the marine equivalents of tropical rain forests, that contain complex interactions among their diverse populations of species.

Open Ocean

The open ocean begins with the sharp increase in water depth at the edge of the continental shelf separates. This open ocean is divided into three vertical zones, based primarily on the penetration of sunlight. The brightly lit upper zone is the euphotic zone. This is where drifting phytoplankton carry out about 40% of the world's photosynthetic activity. Nutrient levels are low, and levels of dissolved oxygen are high. Large, fast-swimming predatory fish populate this zone.

The dimly lit middle zone is the bathyal zone. It gets little sunlight, so it does not contain photosynthesizing producers. Zooplankton and smaller fishes, many of which migrate to feed on the surface at night, populate this zone.

The deepest zone is called the abyssal zone. The abyssal zone is dark and very cold. It has little dissolved oxygen, but is teeming with life because it contains enough nutrients to support a large number of species, even though there is no sunlight to support photosynthesis.

Freshwater Systems

Freshwater life zones include lakes, ponds, inland wetlands (lentic), and streams and rivers (lotic). Although these freshwater systems cover less than 2% of the earth's surface, they do provide important ecological and economic services.

Lakes are large bodies of standing freshwater formed when precipitation, runoff, or groundwater fills depressions in the earth's surface. These depressions can be caused by glaciers, displacement of the earth's crust, or volcanic activity. Lakes are supplied with water from rainfall, melting snow, and streams that drain their surrounding watershed.

Lakes that are deep consist of four zones which are defined by their depth and distance from shore. The top layer is called the littoral zone. This zone is near the shore and consists of shallow sunlit waters to depth at which plants stop growing. It is biologically diverse because of ample sunlight and nutrients from the surrounding land.

The next layer is the limnetic zone. This is the open, sunlit surface layer away from the shore that extends to the depth penetrated by sunlight. This is the main photosynthetic body of the lake and produces the food and oxygen that support the lake.

The profundal zone is the deep, open water where it's too dark for photosynthesis to occur. Without sunlight and plants, oxygen levels are low. Fishes adapted to the lake's cooler and darker water are found in this zone.

The bottom of the lake contains the benthic zone. It's inhabited mainly by decomposers, detritus feeders, and some fishes. The benthic zone is nourished mainly by dead matter that falls from both the littoral and limnetic zones.

Module 10 Goals

The following is a list of goals that should be met by the end of this module. The goals are very broad and, sometimes, general in nature. This was done to allow you to decide how much depth you'll want to go into for each goal.

Examining the Progress Check questions will also help focus your attention better on the goals.

1. Describe the general nature of aquatic systems.
2. Differentiate aquatic and marine ecosystems.
3. Explain the importance of marine aquatic systems.
4. Explain how human activities have affected marine ecosystems.
5. Explain how human activities have affected freshwater ecosystems.

Module 10 Key Words

Here's a list of key words for this module:

aquatic life zones
benthos
coastal wetlands
estuaries
eutrophic lake
freshwater systems
hypereutrophic lakes
intertidal zone
lakes
marine systems
mesotrophic lakes
nekton
oligotrophic lakes
open sea
plankton
turbidity
watershed

Module 10 Progress Check

To check your understanding of the goals of this module, you should be able to answer the following questions.

1. What percentage of the earth's surface is covered with water?
2. Distinguish between a marine life zone and a freshwater life zone.
3. Define nekton, benthos, and decomposers and give an example of each.
4. What five factors determine the types and numbers of organisms found in the three layers of aquatic life zones?
5. What major ecological and economic services are provided by marine systems?
6. Distinguish among oligotrophic, eutrophic, hypereutrophic, and mesotrophic lakes.
7. What is cultural eutrophication?
8. Describe the relationships between dams, deltas, wetlands, hurricanes, and flooding in New Orleans, Louisiana.
9. What are four ways in which human activities are disrupting and degrading freshwater systems?
10. How is the degradation of many of the earth's coral reefs a reflection of our failure to follow the four scientific principles of sustainability?

Module 11 Maintaining Aquatic Biodiversity

Module 11 Overview

It's been said that we have explored outer space than we have the oceans. We have only explored about 5% of the earth's oceans and know very little about its biodiversity. We also have limited knowledge about freshwater biodiversity. But what we do know is this: the greatest marine biodiversity occurs in coral reefs, estuaries, and the deep-ocean floor; biodiversity is higher near coasts than in the open sea; and biodiversity is higher in the bottom region of the ocean than in the surface region.

One major threat is the loss and degradation of many sea-bottom habitats due to dredging operations and trawler fishing boats. Trawlers drag huge nets weighted down with heavy chains and steel plates like giant submerged bulldozers over ocean bottoms to harvest a few species of bottom fish and shellfish. These trawling nets destroy coral reef habitats and kill a wide variety of life on the bottom. Thousands of trawlers scrape and disturb an area of ocean floor about 150 times larger than the area of forests that are clear-cut annually.

Habitat disruption is also a problem in freshwater zones. Dams and excessive water withdrawal from rivers and lakes destroy aquatic habitats and disrupt freshwater biodiversity. As a result of these activities, it's estimated that 51% of freshwater fish species are threatened with extinction.

Another problem is the deliberate and accidental introduction of hundreds of harmful invasive species into coastal waters, wetlands, and lakes throughout the world. These invaders displace and cause the extinction of native species and disrupt the ecosystem. In the late 1980s, Lake Victoria was invaded by the water hyacinth. This rapidly growing plant has carpeted large areas of the lake, blocked sunlight, deprived fish and plankton of oxygen, and reduced aquatic plant diversity.

People also introduce invasive species. The Asian swamp eel has invaded the waterways of south Florida because of the dumping of home aquariums. This eel rapidly reproduces and eats almost anything by sucking them in like a vacuum cleaner. It eludes cold weather, drought, and predators by burrowing into mud banks. It's resistant to waterborne poisons because it can breathe air. It can wriggle across dry land to invade new waterways, ditches, canals, and marshes. It's entirely possible that this eel could take over much of the waterways of the southeastern United States.

Toxic pollutants from industrial and urban areas kills aquatic life by poisoning them. Each year plastic items dumped from ships and left as litter on beaches kill up to 1 million seabirds and 100,000 mammals and sea turtles. Such pollutants and debris threaten the lives of millions of marine mammals and countless fish that ingest, become entangled in, or are poisoned by them. These forms of pollution lead to an overall reduction in aquatic biodiversity and degradation of marine ecosystems.

Climate change also threatens aquatic biodiversity and ecosystems by causing sea levels to rise. During the past 100 years, average sea levels have risen by 4–8 inches. Scientists estimate they will rise another 0.6–1.9 feet and perhaps as high as 3.2–5.2 feet between 2050 and 2100. This would destroy more coral reefs, swamp some low-lying islands, drown many highly productive coastal wetlands, and put much of the U.S. state of Louisiana's coast, including New Orleans, under water.

The human demand for seafood is outgrowing the sustainable yield of most ocean fisheries. To keep consuming seafood at the current rate, it's estimated we'll need 2.5 times the area of the earth's current oceans. Overfishing is not new. Archaeological evidence shows that for thousands of years, humans living in coastal areas have over-harvested fishes, shellfish, seals, turtles, whales, and other marine mammals. But today's industrialized fishing fleets can overfish much more of the oceans and deplete marine life at a much faster rate.

Protecting marine biodiversity is difficult for a number of reasons. First, human ecological footprint and fishprints are expanding so rapidly that it's difficult to monitor the impacts. Second, much of the damage to the oceans and other bodies of water isn't visible to most people. Third, many people wrongly view the oceans as an inexhaustible resource able to absorb an infinite amount of waste and pollution and still produce all the seafood we want. Lastly, most of the world's oceans lie outside the legal jurisdiction of any country. It's an open-access resource, subject to overexploitation.

National and international laws and treaties have been enacted to help protect marine species. These regulations include the 1975 Convention on International Trade in Endangered Species, the 1979 Global Treaty on Migratory Species, the U.S. Marine Mammal Protection Act of 1972, the U.S. Endangered Species Act of 1973, the U.S. Whale Conservation and Protection Act of 1976, and the 1995 International Convention on Biological Diversity.

Module 11 Goals

The following is a list of goals that should be met by the end of this module. The goals are very broad and, sometimes, general in nature. This was done to allow you to decide how much depth you'll want to go into for each goal.

Examining the Progress Check questions will also help focus your attention better on the goals.

1. Identify the major threats to aquatic biodiversity.
2. Describe how we can protect and sustain marine biodiversity.
3. Define how we should manage and sustain marine fisheries.
4. Explain how we can protect and sustain wetlands.
5. Explain how we protect and sustain freshwater lakes, rivers, and fisheries.
6. Identify our priorities for sustaining biodiversity and ecosystem services.

Module 11 Key Words

Here's a list of key words for this module:

fishprint
maximum sustained yield
mitigation banking
multispecies management
optimum sustained yield
precautionary principle

Module 11 Progress Check

To check your understanding of the goals of this module, you should be able to answer the following questions.

1. What are three general patterns of marine biodiversity?
2. Why is marine biodiversity higher (a) near coasts than in the open sea and (b) on the ocean's bottom than at its surface?
3. Describe the threat to marine biodiversity from bottom trawling.
4. Give two examples of threats to aquatic systems from invasive species and explain why they are threats.
5. What is a fishprint and why is it significant?
6. Describe the international efforts to protect whales from overfishing and premature extinction.
7. Describe the major threats to the world's rivers and other freshwater systems.
8. What are three major ecological services provided by wetlands?
9. Describe invasions of the U.S. Great Lakes by nonnative species.
10. What are the six priorities for protecting terrestrial and aquatic biodiversity?

Unit 4 - Earth's Resources

Module 12 Food, Soil and Pests

Module 12 Overview

There is more than enough food produced today to meet the basic nutritional needs of every person on the earth. Even with this food surplus, one of every six people in developing countries is not getting enough to eat. These people face what is known as food insecurity; living with chronic hunger and poor nutrition, which threatens their ability to lead healthy and productive lives.

Experts agree that the cause of food insecurity is poverty. Since 1990, India has produced enough grain to feed its entire population. But about one-fifth of the country's population are hungry because they cannot afford to buy or grow enough food.

Other obstacles to food security include political instability, corruption, and war. These interfere with the distribution and transportation of food and result in people going hungry while stored foods spoil or are distributed unevenly. Achieving food security on regional and global levels for both poor and affluent people also depends on greatly reducing the harmful environmental effects of agriculture.

Food Production

Three systems supply most of our food. Croplands produce mostly grains and provide about 77% of the world's food using 11% of its land area. Rangelands, pastures, and feedlots produce meat and supply about 16% of the world's food using about 29% the world's land area. Oceanic fisheries, and more recently aquaculture, supply about 7% of the world's food.

These three systems depend on a small number of plant and animal species. Of the estimated 50,000 plant species that people can eat, only 14 of the these supply an estimated 90% of the world's food calories. Wheat, rice, and corn provide about 47% of the calories and 42% of the protein people consume and two-thirds of the world's population survive on these three grains. This food specialization puts us in a vulnerable position if the small number of crop strains, livestock breeds, and fish species we fail because of disease, environmental degradation, and climate change.

Since 1960, there has been a tremendous increase in global food production from all three of the major food production systems. This happened because of technological advances like increased use of tractors and farm machinery and high-tech fishing equipment. Other developments include chemical fertilizers, irrigation, pesticides, high-yield grain, and raising large numbers of livestock, poultry, and fish in factory-like conditions. Farmers can produce more food either by farming more land or by getting higher yields from existing land. Since 1950, 88% of the increase in global food production has come from using industrialized agriculture to increase crop yields in what is called the green revolution.

The green revolution consists of three steps. The first step is to develop selectively bred or genetically engineered high-yield crops like rice, wheat, and corn. The second step is to produce high yields by using large amounts of fertilizers, pesticides, and water. The third, and last step, is to increase the number of crops grown each year on a piece of land through multiple cropping. Between 1950 and 1970, this high-input approach dramatically increased crop yields in what was called the first green revolution.

A second green revolution has been taking place since 1967. Fast-growing dwarf varieties of rice and wheat, specially bred for tropical and subtropical climates, have been introduced in India, China, and several developing countries in Central and South America. Producing more food on less land has the benefit of protecting biodiversity by saving large areas of forests, grasslands, wetlands, and easily eroded mountain terrain from being used to grow food crops. Because of these two green revolutions, world grain production tripled between 1950 and 1996. Per capita food production increased by 31% between 1961 and 1985, but has generally declined since.

In the United States, industrialized farming evolved into agribusiness, with a small number of giant multinational corporations controlling the growing, processing, distribution, and sale of food in the United States and global marketplaces. Agriculture is bigger than the country's automotive, steel, and housing industries combined In total annual sales. It generates one-fifth of the country's gross domestic product. U.S. farms produce about 17% of the world's grain with only 0.3% of the world's farm labor force.

Since 1950, industrialized agriculture has more than doubled the yields of wheat, corn, and soybeans without cultivating more land. These yield increases have kept large areas of forests, grasslands, and wetlands from being converted to farmland. Farmers have used crossbreeding through artificial selection to develop genetically improved varieties of crops and livestock animals for centuries. This type of selective breeding is the first gene revolution and has yielded amazing results. The potato discovered high in the Andes mountains by the Spanish look nothing like the potatoes eaten today.

But this crossbreeding is a slow process It can take 15 or more years to produce a commercially valuable crop variety, and can only combine traits from species that are genetically similar. The resulting varieties remain useful for only 5–10 years before pests and diseases reduce their effectiveness.

Scientists have created a second gene revolution by using genetic engineering to develop genetically improved strains of crops and livestock animals. Genetic engineering involves altering an organism's genetic material through adding, deleting, or changing segments of its DNA, to produce desirable traits or to eliminate undesirable ones. It enables scientists to transfer genes between different species that would not interbreed in nature. The resulting organisms are called genetically modified organisms (GMOs). Genetic engineers have used genes from ordinary daffodils and a soil bacterium to produce golden rice. Compared to traditional crossbreeding, genetic engineering of this new variety of rice took about half as long to develop.

Ready or not, the world is entering the age of genetic engineering. Bioengineers are developing new varieties of crops that are resistant to heat, cold, herbicides, insect pests, parasites, viral diseases, drought, and salty or acidic soil. They also have the ability to develop crops that can grow faster, survive with little or no irrigation, with less fertilizer, and no pesticides.

The downside to all this good news is that agriculture has a greater harmful environmental impact than any human activity and these environmental effects may limit future food production. It's possible that crop yields in may decline because of erosion, degradation of soil, and pollution of underground water supplies used for irrigation. According to the U.S. Environmental Protection Agency, agriculture is responsible for three quarters of the water quality problems in U.S. rivers and streams.

Some of the impacts of agriculture include:

- topsoil erosion
- excessive irrigation
- huge inputs of energy
- controversy over GMO's
- loss of biodiversity

Pests

The biggest problem farmers have historically faced have been pests. The control of agricultural pests to increase crop yields is an ongoing battle faced every year by farmers around the world. Of course we didn't invent the use of chemicals to kill or repel pests. Plants have been producing their own insecticides for over 225 million years.

The first artificially produced insecticide that was highly potent was DDT. DDT had been known since 1874 but in 1939 Paul Muller discovered that it was a very potent insecticide. It quickly became the world's most widely used insecticide. In 1962 Rachel Carson sounded a warning about the dangers of DDT that eventually led to the banning of DDT use around the world.

Since then the development and use of insecticides around the world has exploded. For example, each year the United States uses about 2.6 million tons of pesticides. They consist of 600 pest-killing chemicals mixed with 1,200 solvents, preservatives, and other supposedly inactive ingredients in about 25,000 commercial pesticide products. Even though the EPA, USDA, and FDA are supposed to regulate the types of insecticides used, the level of exposure, and the health risks involved, these chemicals are in the run-off where they enter the soil and eventually the water table.

Whatever isn't washed off our fruits and vegetables in the process of going from farm to table, we ingest. If you want to see a list of the worst and best foods, go to Inspiration Green at www.inspirationgreen.com. What you'll find will be very thought provoking.

The problem with overuse of artificial insecticides is one of resistance. Insects, and other pests, adapt very easily to chemical insecticides. The more they become resistant, the stronger and more powerful the insecticide. The more powerful the insecticide, the more resistant the pests become. It can become a never ending upward spiral. The optimal solution is to incorporate more natural alternatives to deal with the pests. Around the world there have been many effective methods developed to protect crops from pests and increase crop yields.

Module 12 Goals

The following is a list of goals that should be met by the end of this module. The goals are very broad and, sometimes, general in nature. This was done to allow you to decide how much depth you'll want to go into for each goal.

Examining the Progress Check questions will also help focus your attention better on the goals.

1. Explain what food security is and why it's difficult to attain.
2. Describe the environmental problems that arise from food production.
3. Describe how we can protect crops from pests more sustainably.
4. Define how we can improve food security.
5. Explain how we can produce food more sustainably.

Module 12 Key Words

Here's a list of key words for this module:

aquaculture
chronic malnutrition
chronic undernutrition
desertification
famine
fishery
food insecurity
green revolution
industrialized agriculture
integrated pest management
overnutrition
pest
pesticides
polyculture
salinization
slash-and-burn agriculture
soil conservation
soil erosion
waterlogging

Module 12 Progress Check

To check your understanding of the goals of this module, you should be able to answer the following questions.

1. Define food security and food insecurity.
2. What is the root cause of food insecurity?
3. Distinguish between industrialized agriculture, plantation agriculture, traditional subsistence agriculture, traditional intensive agriculture, polyculture, and slash-and-burn agriculture.
4. Describe formation of soil and the major layers in mature soils.
5. What are the major harmful environmental impacts of agriculture?
6. What is desertification and what are its harmful environmental effects?
7. Explain how most food production systems reduce biodiversity.
8. What are the seven alternatives to conventional pesticides?
9. How can we make aquaculture more sustainable?
10. Define organic agriculture and describe its advantages over conventional agriculture.

Module 13 Fresh Water Resources

Module 13 Overview

Water is an amazing substance with unique properties that affects all life on earth. Water covers about 71% of the earth's surface. Organisms are made up of about 60% water, most of it inside cells. Humans can survive for several weeks without food but for only a few days without water. It takes huge amounts of water to supply the population with food, shelter, and meeting other daily needs and. Water plays a key role in sculpting the earth's surface, moderating climate, and removing or diluting wastes and pollutants. In spite of all this, water is one of the most poorly managed resources. It is wasted, polluted, and is charged too little for its availability. This increases still greater waste and pollution of this resource, for which there is no substitute.

Access to useable water is a global health issue because lack of water that is safe is the world's single largest cause of illness. The World Health Organization estimates that more than 1.5 million children under age five die from preventable waterborne diseases such as typhoid fever, and hepatitis.

Water is an economic issue because it is vital for reducing poverty and producing food and energy. In developing countries it's a women's and children's issue because poor women often are responsible for finding daily supplies of water. Water is a national and global security issue because of increasing tensions within and between nations over access to limited but shared water resources in the Middle East and other areas of the world. One of the scenarios with overpopulation is one which involves conflict over sources of useable water. As water supplies dwindle and population goes up, especially in those areas of the world where food and water is already scarce, where political stability is a day-to-day thing, these areas of the world are ripe for major conflict.

As an environmental issue the excessive withdrawal of water from rivers and aquifers as well as the pollution of water results in lower water tables, lower river flows, shrinking lakes, losses of wetlands, declining water quality, declining fish populations, species extinctions, and destruction of ecosystems in aquatic systems.

The earth has plenty of water, but only a fraction of it (about 0.024%) is available as freshwater in accessible deposits. The rest is in the salty oceans, frozen in polar ice caps and glaciers, or deep underground. Fortunately, the world's freshwater supply is continually collected, purified, recycled, and distributed through the water cycle. The water recycling and purification system works well, unless it's overload with slowly degradable and non-degradable wastes or withdrawn from underground and surface water supplies faster than it's replenished. The water cycle is also interfered with when wetlands are destroyed, or forests are cut down that store and slowly release water, or cycle's rate and distribution patterns are altered as a result of climate change.

The good news is that the world has plenty of freshwater. Unfortunately differences in average annual precipitation and economic resources divides countries and people into water haves and have-nots. For example, Canada, with only 0.5% of the world's population, has 20% of the world's freshwater. China, on the other hand, has 20% of the world's population with only 7% of the water supply.

The United States has more than enough renewable freshwater, but it is unevenly distributed, and a lot of it is contaminated through agricultural and industrial practices. The eastern states usually get a lot of precipitation, while many western and southwestern states get little.

In the East, most of the water is used for energy production, power plant cooling, and manufacturing. In parts of the eastern United States, the most serious water problems are flooding, occasional urban shortages as a result of pollution, and drought. In the western half of the United States, irrigation counts for 85% of water use, much of it unnecessarily wasted. The major water problem is a shortage of runoff because of low precipitation, high evaporation, and recurring severe drought.

About half of the water used in the United States comes from groundwater sources. The rest comes from rivers, lakes, and reservoirs. Water tables in many water-short areas are dropping too quickly as farmers and rapidly growing urban areas deplete the aquifers faster than they can be recharged. Excessive withdrawal of groundwater from an aquifer near a river or stream can deplete these sources of surface water. For example, when the water table drops below the level of the bottom of a stream, the stream water drains into the aquifer.

In 2007, the U.S. Geological Survey projected that approximately 36 states were going to face water shortages by 2013 because of a combination of drought, rising temperatures, population growth, urban sprawl, and excessive use of water. Scan the news headlines and you can see how much of this prediction is occurring. In 2003, the U.S. Department of the Interior mapped out water hotspots in 17 western states. In these areas, competition for scarce water would likely trigger intense political and legal conflicts between states and also between rural and urban areas within states during the next 20 years.

This is already happening in a case in which Montana accused Wyoming of taking more than its fair share of water from two tributaries of the Yellowstone River that supply water for wells and farms in both states. Predicted long-term drought and global warming will increase the number and intensity of these kinds of disputes in the U.S. and around the world in the years to come.

In the United States, the Ogallala aquifer supplies about one-third of all the groundwater used in the U. S. and is responsible for turning the Great Plains into one of world's most productive agricultural regions. The problem is that the Ogallala is a one-time deposit of water with a very slow rate of recharge.

In some areas of the Great Plains water is being pumped out at a rate 10–40 times higher than the natural recharge rate. This has lowered the water table more than 100 feet and has pumping costs to increase making it too expensive to irrigate crops in some areas. For example, in recent years the amount of irrigated farmland in Texas has decreased by about 11%, since farmers give up agriculture or switch to lower yield dryland farming. Added to this problem are the government subsidies designed to increase crop production and encourage farmers to grow water-thirsty crops. This policy has accelerated the depletion of the Ogallala.

The global solution is to find ways of using water more sustainably. This can be done through programs that help by cutting water waste, by raising water prices, slowing population growth, and protecting the aquifers, forests, and other ecosystems that store and release water.

Module 13 Goals

The following is a list of goals that should be met by the end of this module. The goals are very broad and, sometimes, general in nature. This was done to allow you to decide how much depth you'll want to go into for each goal.

Examining the Progress Check questions will also help focus your attention better on the goals.

1. Examine the limits to usable water globally.
2. Discuss how extracting groundwater, building more dams, transferring water from one place to another, and converting salty seawater to freshwater are possible answers to decreasing usable water supplies.
3. Define how we can use water more sustainably.

Module 13 Key Words

Here's a list of key words for this module:

aquifiers
drainage basin
drought
floodplain
groundwater
surface runoff
surface water
zone of saturation

Module 13 Progress Check

To check your understanding of the goals of this module, you should be able to answer the following questions.

1. What percentage of the earth's freshwater is available to us?
2. Define groundwater, zone of saturation, water table, and aquifer.
3. Define surface water, surface runoff, and watershed.
4. Distinguish between surface runoff and reliable surface runoff.
5. Who should own and manage freshwater resources and why?
6. Describe the problem of groundwater depletion in the world and in the United States, especially over the Ogallala aquifer.
7. Define desalination and distinguish between distillation and reverse osmosis as methods for desalinating water.
8. What are the limitations of desalination and how might they be overcome?
9. What percentage of the world's water is unnecessarily wasted and what are two causes of such waste?
10. List three human activities that increase the risk of flooding.

Module 14 Geology and Minerals

Module 14 Overview

No consideration of environmental science is complete without examining the earth's geology and the mineral resources hidden below its surface.

Current estimates put the age of the earth at around 4.5 billion years. Life made its first appearance around 3.5 billion years ago. In all that time the surface of the earth has undergone major transformations and will continue to do so for many millions of years to come.

The earth is constantly undergoing a change, however slow it is. Modern geology is based on this idea, referred to as uniformitarianism, that geologic change occurs over many millions of years. It's ironic that in his early travels on the H.M.S. Beagle, Charles Darwin familiarized himself with the work of Charles Lyell, England's leading geologist and proponent of uniformitarianism. This idea of change over time became a central theme in Darwin's theory of evolution.

Along with the biological and chemical cycles, there is also a geological cycle that is most often referred to as the rock cycle. What we call the land is actually the upper surface of huge tectonic plates. There are a total of 15 tectonic plates that make up the crust of planet. They literally float on top of the mantle. The mantle, simply put, represents enormous convection currents that carry heat from the core to the surface.

The plates are constantly in motion and every 300-500 million years all the continents come together to form a supercontinent. The last supercontinent, Pangea, occurred about 300 million years ago. And we've been on the move ever since. The North American Plate, for example, is moving in a southwest direction at about an inch a year.

The movement of the plates isn't so much a problem as where the plates meet. All 15 plates are touching each other and the place where they meet is called the plate boundary. This is where the problems occur in environmental science. There are three types of plate boundaries; transform, divergent, and convergent. There are a tremendous forces in place at these boundaries and when the plates overcome the forces of friction bad things happen.

Transform boundaries occur where one plate is literally sliding another. The San Andreas Fault is an example. The Pacific plate is generally moving in a northerly direction. It's eastern border follows the coastline of California, British Columbia, and Alaska. The North American Plate is moving generally in a southerly direction. Since there is a great deal of friction between the two, it's only a matter of time when the forces of friction are overcome by the movement of the plates and they move, or more accurately, snap. The resulting earthquake can be minor or major, causing billions of dollars worth of damage.

Divergent boundaries occur where the plates move away from each other. These boundaries are not marked by earthquakes so much as they are by volcanoes. The mid-Atlantic ridge is an example of this type of boundary along with the East African Rift where the Indian plate and the African plate are moving away from each other.

Convergent boundaries occur where plates move into each other. When that happens one of two event occur. If the two plates move into each other and the crust of the earth is pushed up, you get mountains. The Himalayas are a great example. The Indian plate is pushing north into the Eurasian plate, which is pushing down. Where these two plates meet are the Himalayan Mountains.

The opposite boundary can occur, where the plates push down on each other in what is known as subduction. Many times volcanoes will form at these boundaries, as they form along all boundaries. Earthquakes created by these faults art just as powerful and dangerous as any other. But they can be terribly destructive when they occur in the ocean. Tsunamis are the result of a great deal of water is pushed into motion as a result of an underwater earthquake.

In March of 2011 a magnitude 9.0 earthquake occurred off the coast of Japan. The resulting tsunami killed close to 20,000 people, destroyed over 129,000 building, severely damaging another 250,000 buildings along with the Fukushima Nuclear Power Plant, and caused over $35 billion in damages.

Earth's volcanoes occur all along plate boundaries. What is known as the Ring of Fire are the series of volcanoes that ring the entire Pacific Ocean. Volcanoes come in different varieties. There are the ones that are found in the Hawaiian Islands. Kilauea began erupting in 1983 and hasn't stopped since. The lava flows directly into the ocean and causes very little damage.

Mauna Loa is the world's largest volcano with a total height from the base to the summit of over 10 miles. It covers 50% of the island of Hawaii and has erupted 33 times since the islands were first settled in 1843. These volcanoes are island builders. They are slow and steady eruptions that add to the land masses.

The volcanoes that cause the most damage are the ones that erupt violently. Mt. Vesuvius in Italy, Mt. St. Helens in Washington, Mt. Pinatubo in the Philippines are examples of these very dangerous and very destructive volcanoes. Probably the most famous of these kind of volcanoes is Krakatoa. Found in the Java Sea, Krakatoa exploded in 1883 so violently that the explosion was heard as far away as 3,000 miles. It caused numerous destructive tsunamis and killed over 37,000 people. It also changed global weather patterns for years as a result of all the ash that was released into the atmosphere. Global temperatures fell by over 1.2 degrees centigrade.

It doesn't take much to imagine the ecological destruction these natural disasters cause through loss of habitat, ecosystem degradation, and local extinction of species. We have no control over these geological processes, other than to maybe not live in areas affected by the movement of the plates.

As damaging as these geologic events are, the earth recovers from them. What has far more reaching consequences is what we are doing in our efforts to dig the riches found below the surface of the planet. The way that we mine for coal, ores, and minerals has historically caused many environmental problems and created much controversy among communities and governments.

Mining, whether it be for coal or gold, consists of a number of different methods, all of which can be bad for the environment if not regulated properly. These methods are surface mining, strip mining, contour mining, subsurface mining, and mountaintop removal. Without strict regulation, and enforcement of these regulations, mining causes loss of habitat, deforestation, destruction of rivers and streams, poisoning of soil and water supplies, erosion, production of solid and liquid hazardous wastes, and leaves the land basically useless. When mining operations shut down, they destroy local economies creating job loss and economic hardship for those left behind.

The earth's crust contains abundant deposits of nonrenewable mineral resources such as iron and aluminum. But mineral resources such as manganese, chromium, cobalt, and platinum are somewhat scarce. Geologic processes have not evenly distributed deposits of mineral resources. Some countries have rich mineral deposits and others have few or none.

Today five nations, the United States, Canada, Russia, South Africa, and Australia, supply most of the nonrenewable mineral resources used by modern societies. South Africa, for example, is nearly self-sufficient in the world's key mineral resources and is the world's largest producer of gold, chromium, and platinum. Three countries, the United States, Germany, and Russia, with only 8% of the world's population consume about 75% of the most widely used metals.

The future supply of nonrenewable minerals depends on two factors: the actual supply of the mineral and the rate at which we use it. We never completely run out of any mineral. A mineral becomes economically depleted when it costs more than it is worth to find, extract, transport, and process. There are five choices then: recycle or reuse existing supplies, waste less, use less, find a substitute, or do without.

Geology determines the quantity and location of a mineral resource. Economics determines what part of the mineral resource is extracted and used. An increase in the price of a scarce mineral resource can lead to increased supplies and can encourage more efficient use, but there are limits to this process.

According to economics, in a competitive market system a plentiful mineral resource is cheap when its supply exceeds demand. When a resource becomes scarce, its price rises. This is supposed to encourage exploration for new deposits, stimulate the development of better mining technology, and make it profitable to mine lower-grade ores. It can also encourage a search for substitutes and promote resource conservation.

But some economists say this pricing model no longer applies very well in most developed countries. Subsidies, taxes, regulations, and import tariffs are often used by governments to control the supplies, demands, and prices of these minerals to such an extent that a competitive market doesn't exist. For example, between 1982 and 2007, U.S. mining companies received more than $6 billion in government subsidies.

Module 14 Goals

The following is a list of goals that should be met by the end of this module. The goals are very broad and, sometimes, general in nature. This was done to allow you to decide how much depth you'll want to go into for each goal.

Examining the Progress Check questions will also help focus your attention better on the goals.

1. Describe the earth's major geological processes and hazards.
2. Explain the earth's rock cycle.
3. Describe mineral resources and what the environmental effects of using them are.
4. Define how long supplies of nonrenewable mineral resources will last.
5. Explain how we can use mineral resources more sustainably.

Module 14 Key Words

Here's a list of key words for this module:

contour mining
core
crust
depletion time
earthquake
geology
igneous rock
lithosphere
mantle
metamorphic rock
mining
ore
rock cycle
sedimentary rock
smelting
strip mining
tectonic plates
tsunami
volcano
weathering

Module 14 Progress Check

To check your understanding of the goals of this module, you should be able to answer the following questions.

1. Define the terms geology, core, mantle, crust, tectonic plate, and lithosphere.
2. Define volcano and describe the nature and effects of a volcanic eruption.
3. Define and describe the nature and effects of an earthquake.
4. What is a tsunami and describe its devastating effects.
5. Describe the nature and importance of the rock cycle.
6. Discuss the major harmful environmental effects of extracting, processing, and using nonrenewable mineral resources.
7. What are five possible options when a mineral becomes economically depleted?
8. What are the pros and cons of the U.S. General Mining Law of 1872?
9. How dependent is the United States on other countries for important nonrenewable mineral resources?
10. How can we use nonrenewable mineral resources more sustainably?

Module 15 Nonrenewable Energy

Module 15 Overview

The term nonrenewable energy refers to the use of coal, petroleum, and natural gas as a source of energy used by the industrialized countries of the world. About 82% of the global energy used comes from these nonrenewable sources; 76% from coal, petroleum and natural gas and 6% from nuclear power. The other 18% comes from renewable sources like solar power, wind turbines, geothermal sources, and hydropower.

Coal, petroleum, and natural gas are called nonrenewable because once they're gone, they're gone. Developing renewable sources of energy has been a major emphasis for many years. But at this point in time, nothing come close to coal, oil, and natural gas in price and efficiency, the environmental problems they create notwithstanding.

There's no question that the world demand for oil has been increasing steadily for decades. China, for example, has increased its demand for oil simple because more and more Chinese are buying cars. They question is how much oil is left and when will it run out. According to analysts, about 75% of the world's oil reserves are in the hands of governments. The rest are privately owned. The 13 countries of OPEC control 60% of the global oil reserves. Natural gas deposits are usually found where there is oil.

The problems with oil not only run from the pollution caused by cars, truck and buses, but the damage done to the environment from pumping and transporting oil. Cars are among the biggest contributors to carbon dioxide emissions. It's a byproduct of burning gas. As a greenhouse gas, carbon dioxide is one contributor to climate change. As the demand for cars goes up, as the human population increases, dealing with carbon emissions from cars is going to remain an ongoing problem until auto manufacturers develop the technology to lower or eliminate carbon emissions.

The other problem with oil is at the production end. Some oil deposits are found in environments that are fragile to begin with, like the tundra of Russia, Alaska, and Canada. Bringing in heavy equipment to drill, extract, and transport the oil safely has always been a concern. The Alaska pipeline runs from the Alaska North Slope to Valdez, a distance of 800 miles. Constructing and maintaining the pipeline has had a major impact on the fragile tundra biomes the pipeline passes through. The damage caused by a break in the pipeline scares environmentalists and oil executives alike.

One of the most famous oil spills in history occurred in 1989 at the end of the pipeline. The tanker Exxon Valdez, carrying 1.48 million barrels of crude oil,ran aground in Prince William Sound and spilled anywhere from 250,000 to 750,000 barrels of oil. The resulting law suits held Exxon liable for the damages and they paid billions of dollars for the cleanup and restoration of the habitats affected. It's estimated that some of those habitats will not be fully restored for years to come.

Natural gas is more plentiful than oil, has a high energy yield, a fairly low cost, and has the lowest environmental impact of all fossil fuels. Natural gas is a mixture of gases of which methane is the most predominant. It contains smaller amounts of heavier gaseous hydrocarbons such as ethane, propane, and butane, and small amounts of hydrogen sulfide.

Natural gas is found above reservoirs of crude oil. Unless a gas pipeline has been built, these deposits cannot be used. The natural gas found above oil reservoirs in deep-sea and remote land areas is often an unwanted by-product and is burned off. This wastes a valuable energy resource and releases carbon dioxide into the atmosphere.

When an oil field is tapped, propane and butane gases are liquefied and removed as liquefied petroleum gas (LPG). LPG is stored in pressurized tanks for use mostly in rural areas not served by natural gas pipelines. The rest of the methane is dried to remove water vapor, cleansed of poisonous hydrogen sulfide and other impurities, and pumped into pressurized pipelines for distribution across land areas.

Russia has about 27% of the world's proven natural gas reserves, followed by Iran (15%) and Qatar (14%). The United States has only 3% of the world's proven natural gas reserves but uses about 27% of the world's annual production.

Natural gas is a versatile fuel that can be burned to heat homes, produce hot water, or to produce electricity. It can be used as a fuel for cars with inexpensive engine modifications. In the United States, a pipeline grid delivers natural gas from domestic wells to towns and cities and directly to 60 million American homes.

Natural gas is also used to run medium-sized turbines that produce electricity. These clean-burning turbines have twice the energy efficiency (50–60%) of coal-burning and nuclear power plants (24–35%). They are cheaper to build, require less time to install, and are easier and cheaper to maintain than large-scale coal and nuclear power plants.

As with any fossil fuel, burning natural gas releases carbon dioxide. However, it releases much less carbon dioxide per than coal, conventional oil, or oil from oil sand and oil shale.

Coal is the world's most abundant fossil fuel. According to the U.S. Geological Survey, identified and unidentified global supplies of coal could last for over 214 years. The United States has 25% of the world's proven coal reserves. Russia has 15%, followed by India with 13%, China with 13%, Australia with 8%, and South Africa with 7%.

Coal is the single biggest air polluter and accounts for at least one-fourth of the world's annual carbon dioxide emissions. The burning of coal is one probably one of the most serious environmental problems of this century. Coal is burned in about 2,100 power plants to generate about 40% of the world's electricity. Using a coal-burning power plant is essentially an inefficient way to boil water and produce steam, which is used to spin turbines and produce electricity.

Coal is mostly carbon, but contains small amounts of sulfur, which are released into the air as sulfur dioxide when the coal is burned. Coal also releases large amounts of soot, carbon dioxide, and trace amounts of mercury. Coal-burning power plants account for 25% of the world's emissions of carbon dioxide and 40% of these emissions in the United States.

China burns a third of the world's coal to provide 70% of its commercial energy. China gets 80% of its electricity from burning coal, and adds the equivalent of three large coal-burning power plants every week. At its current rate of consumption, China has about 37 years of coal reserves left, and only 10–15 years if its coal consumption continues to increase by 10–15% a year.

China and the world are paying a heavy environmental price for its dependence on coal. Pollution controls on older, inefficient plants in China are nonexistent. Even the newest coal-burning plants are inefficient and have inadequate air pollution control systems. Since 2005, China has been the world's leading source of sulfur dioxide, which can cause respiratory and cardiovascular diseases. And sulfur dioxide and nitrogen oxides released by China's coal-burning power plants interact in the atmosphere to form harmful acidic compounds that fall as acid rain in parts of China and other countries. This pollution contributes to the air quality problems in cities such as Seoul, South Korea and Tokyo, Japan.

Because coal has a huge impact on the environment, both in mining it and burning it, there is growing public opposition to building more coal-fired power plants in the United States. In 2007, there were 151 coal-fired power plants in the planning stage. By 2008, 59 of these proposed plants had either been abandoned or refused licenses because the companies planning them were not evaluating less harmful alternative ways to meet the demand for electricity. An additional 50 of the proposed plants are being contested in the courts. Coal will still be burned in many U.S. power plants, but any significant expansion of use of this resource seems unlikely.

The post World War Two hope that nuclear power would solve the world's energy problems have pretty much disappeared. The construction of nuclear power plants is a complex and very expensive undertaking. Running and maintaining the facilities is also complex and expensive. The question of disposing the nuclear waste has never been answered properly and the perceived safety of the plants has been shown to be problematic also.

Two nuclear power plant accidents brought to light the concerns over the safety of these facilities. In 1979 the Three Mile Island power plant outside of Harrisburg, Pennsylvania lost reactor coolant in one of its reactors. The loss of coolant caused part of the radioactive core to become exposed and the high heat associated with radioactive exposure caused the core to collapse to the floor of the reactor. The containment building kept most of the radioactivity inside but some did leak out into the surrounding community. No one knows for sure how much radioactivity was released and the whole incident pretty much brought the construction of nuclear power plants in the U.S. to a halt.

In 1986 the world's most serious nuclear accident happened at the Chernobyl nuclear power plant in the Ukraine. A series of explosions inside the plant, due to poor construction, blew the top off one of the reactors. Huge amounts of radioactivity were released and spread over a wide area of Russia and Europe. There is conflicting data at the eventual death toll as a result of exposure to high levels of radiation. The Russians put the death toll as high as 200,000 while the World Health Organization put it at 9,000. Chernobyl has been isolated due to the high level of radiation in the surrounding countryside. The levels of radiation are expected to remain dangerously high for many, many decades.

Opponents of nuclear power argue that it makes more sense to invest private and government funds in not trying to make nuclear power more safe, but to stimulate the development of energy efficiency and renewable energy resources that are much safer and more easily developed.

Module 15 Goals

The following is a list of goals that should be met by the end of this module. The goals are very broad and, sometimes, general in nature. This was done to allow you to decide how much depth you'll want to go into for each goal.

Examining the Progress Check questions will also help focus your attention better on the goals.

1. Identify the major sources of nonrenewable energy we use.
2. Describe the advantages and disadvantages of each of these: oil, natural gas, coal, and nuclear energy.

Module 15 Key Words

Here's a list of key words for this module:

crude oil
natural gas
petrochemicals
shale oil

Module 15 Progress Check

To check your understanding of the goals of this module, you should be able to answer the following questions.

1. Summarize the issue of whether or not and when we are likely to run out of affordable oil.
2. Explain why the United States cannot even come close to meeting its oil needs by increasing domestic oil supplies.
3. What are the major advantages and disadvantages of using heavy oils produced from oil sand and oil shales as energy resources?
4. What are the major advantages and disadvantages of using natural gas as an energy resource?
5. What are the major advantages and disadvantages of using coal as an energy resource?
6. What is coal liquefaction and how can liquid fuels be produced from coal?
7. What factors have hindered the development of nuclear power?
8. Describe the nuclear power plant accidents at Three Mile Island and Chernobyl.
9. What are the major advantages and disadvantages of relying on nuclear power as a way to produce electricity
10. Summarize the arguments for and against relying more on nuclear power.

Module 16 Renewable Energy

Module 16 Overview

The ideal scenario is to decrease dependence on fossil fuels and increase usage of renewable sources of energy. We can also practice energy-saving behaviors, regardless of the source of the energy.

The best way to conserve energy is to improve energy efficiency, which is the measure of how much work we can get from each unit of energy we use. An obvious example are the people who drive energy-efficient cars who use less fuel per mile than do those who drive less efficient vehicles. Or the adoption of compact fluorescent lightbulbs and the newer LED lightbulbs.

Reducing energy waste has many economic and environmental advantages. Reducing energy waste is the quickest, cleanest, and cheapest way to provide more energy, reduce pollution and environmental problems, slow global warming, and increase economic and national security. The cheapest and cleanest power plant in the world, one that emits no greenhouse gases and produces no radioactive wastes, is the one that isn't built.

Conserving energy can be achieved at all levels of our society with technologies that are available now. From the energy-efficient design of factories, office buildings, and homes, to the way electricity is produced and delivered to the customer. It's estimated that the economic benefits are tremendous, but the costs of getting there are high.

Half of the energy consumed in the U.S. is in transportation. As long as Americans perceive that the cost of a gallon of gas is cheap, they will drive big, energy inefficient cars. There has always been the question of designing a highly efficient gasoline engine that can get 50-60 miles per gallon and how the petroleum interests have prevented the development of such cars.

But progress is being made with the introduction of hybrid and electric cars. The critics of electric cars say that the cost of producing the electricity to charge the car's batteries could end up being higher that the cost of gas. Add to that the increased carbon dioxide emissions from the power plants producing the electricity and you have limited application of an all electric car.

Building construction is another area when progress is being made to conserve energy. Many skyscrapers built today are taking advantage of energy-saving technologies to reduce energy costs. Houses can be super-insulated to use 90% less energy than a traditional house. These homes cost about 5% more to build, but the energy savings alone recoups the increased cost in about 5 years.

Existing homes can be made more energy efficient by doing a number of things. Re-insulating or adding more insulation, checking for and plugging leaks in air conditioning and heating ducts, and around windows. Installing double-paned windows, roof-mounted solar water heaters, and solar electric panels are all technologies that are available now and are in use in many places around the world. Another area that has shown remarkable improvement in energy consumption is in consumer electronics. Everything from appliances to TV's to computers are much more energy efficient than years ago.

Even electrical lighting has undergone a major transformation. A compact fluorescent bulb produces as much light as a regular incandescent bulb, but lasts up to ten times longer and uses one-fourth as much energy, which pays for its higher price in a few months. The U.S. is slowly phasing out the incandescent bulb with the more highly efficient compact fluorescent bulb which will eventually be phased out by the even more energy efficient LED and OLED bulbs. The trend started in the U.S. is slowly spreading to other countries around the world.

Besides solar power, other technologies used to generate power include wind power, geothermal power, hydropower, and biomass. All of these are in use in limited applications and all of these can be used to supplement current energy requirements. But none of these are practical enough to replace fossil fuels and the single source of energy. We may never truly become independent of fossil fuels, but we can get to the point where we are not entirely dependent on fossil fuels for energy.

Module 16 Goals

The following is a list of goals that should be met by the end of this module. The goals are very broad and, sometimes, general in nature. This was done to allow you to decide how much depth you'll want to go into for each goal.

Examining the Progress Check questions will also help focus your attention better on the goals.

1. Describe why energy efficiency is an important energy source.
2. Describe the way we can cut energy waste.
3. Explain the advantages and disadvantages of: solar energy, producing electricity from moving water, producing electricity from wind, biomass as an energy source, geothermal energy, and hydrogen as an energy source.
4. Explain how we can make a transition to a more sustainable energy future.

Module 16 Key Words

Here's a list of key words for this module:

biofuels
energy conservation
energy efficiency
geothermal energy
solar heating

Module 16 Progress Check

To check your understanding of the goals of this module, you should be able to answer the following questions.

1. Distinguish between energy conservation and energy efficiency.
2. Explain why energy efficiency can be thought of as an energy resource.
3. What are the major advantages of reducing energy waste?
4. Describe the trends in fuel efficiency in the United States since the 1970s.
5. Describe five ways to save energy in an existing building.
6. List five advantages of relying more on a variety of renewable sources of energy and describe two factors holding back such a transition.
7. What is a solar cell and what are the major advantages and disadvantages of using such cells to produce electricity
8. What are biofuels and what are the major advantages and disadvantages of using (a) biodiesel and (b) ethanol to power motor vehicles?
9. Evaluate the use of corn, sugar cane, and cellulose plants to produce ethanol.
10. List three general conclusions from energy experts about possible future energy paths for the world.
11. List five major strategies for making the transition to a more sustainable energy future.
12. Describe three roles that governments play in determining which energy resources we use.

Unit 5 - Pollution

Module 17 Environmental Hazards

Module 17 Overview

We face health hazards not only from our lifestyle choices, but also from biological, chemical, physical, and cultural factors. Understanding what these risks are and how we can either control them, or avoid them, will provide give us the tools to lead a healthy life.

A health risk is the probability of suffering harm from a hazard that can cause injury, disease, death, economic loss, or damage. Risks are expressed in terms of probability, which is a statement about how likely an outcome is going to happen. Probabilities are often expressed as ratios or percentages. But there is a difference between possibility and probability. When we say that it's possible that smoking can cause lung cancer, we're saying that this event could happen. The probability of getting ling cancer gives us an estimate of the likelihood of it happening.

Using statistical methods to estimate how much harm a particular hazard can cause to human health or to the environment is called risk assessment. Scientists risk assessment to estimate the probability of a risk, to compare it with the probability of other risks, and to establish priorities for avoiding or managing risks. Risk management involves deciding how to reduce a risk to a certain level and at what cost.

Because the media loves to sensationalize their coverage of any news, most people aren't good at understanding and comparing risks and that becomes a problem. Many people end up worrying about the highly unlikely possibility of minor risks and ignore the significant probability of major risks.

Here's an example. In 2008, because of the sensational coverage the media gave it, people scrambled to get the vaccine for the Avian flu. No one died from the Avian flu in the U.S. In 2008. But these very same people refused to get the vaccine for the common flu which kills about 36,000 people each year.

There are five major types of hazards:

Biological hazards include the more than 1,400 pathogens that can infect humans which include bacteria, viruses, parasites, protozoa, and fungi. Chemical hazards come from harmful chemicals in the air, water, soil, and food. Physical hazards like fire, earthquakes, volcanic eruptions, floods, and storms. Cultural hazards such as unsafe working conditions, unsafe highways, criminal assault, and poverty. Lifestyle choices such as smoking, eating, and drinking too much alcohol, and having unsafe sex.

An infectious disease is caused when a pathogen (bacteria, virus, or parasite) enters the body and multiplies. Examples of infectious diseases include are flu, HIV, malaria, tuberculosis, and measles. A transmissible disease (contagious/communicable disease) is an infectious disease that is transmitted from one person to another. Flu, HIV, tuberculosis, and measles are contagious diseases.

In 1900, infectious disease was the leading cause of death in the world and in the United States. Since the infectious diseases and the death rates from such diseases has been greatly reduced. This has been accomplished by better health care, use of antibiotics, and the development of vaccines. As a result, average life expectancy has increased in most countries, and the leading cause of death has shifted to non-transmissible cardiovascular disease, and cancers. Globally, the biggest biological risk today is from tuberculosis, the flu, HIV, and malaria. More people will die from these four infectious diseases than from any other biological source.

There is growing concern about chemicals that cause birth defects, cancers and disrupt the human immune, nervous, and endocrine systems. A toxic chemical causes temporary or permanent harm to humans and animals. The Environmental Protection Agency lists arsenic, lead, mercury, vinyl chloride, and polychlorinated biphenyls (PCBs) as the five most toxic substances.

Chemically toxic agents fall into one of three categories. Carcinogens are chemicals, types of radiation, or certain viruses that can cause or promote cancer. Arsenic, benzene, chloroform, formaldehyde, gamma radiation, nickel, PCBs, radon, chemicals in tobacco smoke, UV radiation, X-rays, and vinyl chloride are examples of commonly found carcinogens

Mutagens are chemicals and forms of radiation that cause mutations to the DNA molecules found in cells. Nitrous acid, for example, is formed by the digestion of nitrite preservatives in foods and can cause mutations linked to stomach cancer in people.

Teratogens are chemicals that cause birth defects to a fetus or embryo. Ethyl alcohol is a teratogen. Drinking during pregnancy can lead to children with low birth weight and a number of physical, developmental, behavioral, and mental problems. Other teratogens are angel dust, benzene, cadmium, formaldehyde, lead, mercury, mescaline, PCBs, vinyl chloride, and hundreds of other organic compounds.

In terms of the number of premature deaths per year and reduced life span, the greatest risk is poverty. The high death toll resulting from poverty is caused by malnutrition, increased susceptibility to normally nonfatal infectious diseases, and fatal infectious diseases transmitted by drinking unsafe water.

After poverty and gender, the greatest risks of premature death result from lifestyle choices that people make. The best ways to reduce the risk of premature death and serious health problems are to avoid smoking and exposure to smoke, lose excess weight, reduce consumption of foods containing cholesterol and saturated fats, eat a variety of fruits and vegetables, exercise regularly, drink little or no alcohol, avoid excess sunlight, and practice safe sex.

Cigarette smoking is the world's most preventable major cause of premature death among adults. The World Health Organization estimated that tobacco use helped to kill 100 million people during the twentieth century and could kill 1 billion people during this century unless cigarette smoking is greatly reduced. The WHO estimates that each year tobacco contributes to the premature deaths of at least 5.4 million people. By 2030, they estimate this number will increase to more than 8 million. According to the Centers for Disease Control (CDC), smoking kills about 442,000 Americans per year.

Most people are not good at assessing the relative risks from the hazards that surround us. Many people deny the high-risk chances of death, or injury, from voluntary activities they enjoy. The risk of injury or death from motorcycling, smoking, hang gliding, and driving are very high when compared to other activities. And yet, people don't think twice about doing any, or all, of these activities. The most dangerous thing that most people in many countries do each day is to drive or ride in a car.

And yet these very same people may be terrified about the possibility of being killed by a gun, dying of the flu, living near nuclear power plant, getting the West Nile virus, being struck by lightning, or dying in an airplane crash. The probabilities of any of these happening are magnitudes lower than the high-risk activities listed above and yet people are "afraid" of them.

There are five factors that cause people to see a technology or a product as being more or less risky than experts judge it to be. First is fear. Many people fear a new, unknown product or technology more than they do an older, more familiar one. For example, people will have a greater fear of nuclear power plants than of the more familiar and highly polluting coal-fired power plants.

A second factor is the degree of control we have. Most of us have a greater fear of things over which we do not have personal control. Some feel safer driving their own cars over long distances and through bad traffic than traveling the same distance on a plane. Third is whether a risk is catastrophic. We usually are more frightened by news of a single catastrophic accident such as a plane crash than we are of a death from smoking. Fourth, people suffer from optimism bias, which is the belief that risks that apply to other people do not apply to them. The fifth factor is that many of the risky things we do are highly pleasurable and give instant gratification, while the potential harm from such activities comes later.

Module 17 Goals

The following is a list of goals that should be met by the end of this module. The goals are very broad and, sometimes, general in nature. This was done to allow you to decide how much depth you'll want to go into for each goal.

Examining the Progress Check questions will also help focus your attention better on the goals.

1. Describe the sources of the major health hazards we face.
2. Explain the types of biological hazards we face.
3. Explain the types of chemical hazards we face.
4. Illustrate how we can evaluate and deal with chemical hazards.
5. Explain how we perceive risks and how can we avoid the worst of them.

Module 17 Key Words

Here's a list of key words for this module:

carcinogens
infectious disease
mutagens
nontransmissible disease
pathogen
risk
toxicology

Module 17 Progress Check

To check your understanding of the goals of this module, you should be able to answer the following questions.

1. What is the difference between risk, risk assessment, and risk management?
2. Differentiate possibility and probability.
3. What is a pathogen?
4. Give an example of a risk from each of the following: biological hazards, chemical hazards, physical hazards, cultural hazards, and lifestyle choices.
5. What are the differences between a non-transmissible disease, infectious disease, and transmissible disease and give an example of each?
6. Give three examples of problems being studied in the new field of ecological medicine.
7. Discuss the use of pollution prevention and the precautionary principle in dealing with health threats from chemicals.
8. What five factors can cause people to misjudge risks?
9. List five principles that can help us evaluate and reduce risk.
10. Discuss how lessening the threats of HIV/AIDS and other major infectious diseases can be achieved by applying the four scientific principles of sustainability.

Module 18 Solid and Hazardous Waste

Module 18 Overview

One of three major categories of waste is solid waste. This is any unwanted or discarded material produced that isn't a liquid or a gas. Solid waste is divided into two types. Industrial solid waste is produced by mines, agriculture, and industries that supply people with goods and services. Municipal solid waste consists of the solid waste produced by homes and workplaces. In developed countries, most municipal solid waste is buried in landfills or burned in incinerators. In many developing countries much of it ends up in open dumps.

Another major category of waste is hazardous, or toxic, waste. This waste threatens human health because it's poisonous, chemically reactive, corrosive, or flammable. Two classes of hazardous wastes are organic compounds and non-degradable toxic heavy metals.

The last category of hazardous waste is radioactive waste produced by nuclear power plants and nuclear weapons facilities. These wastes have to be stored safely for 10,000 to 240,000 years depending on the radioactive isotopes present. After 60 years of research, there is still considerable scientific disagreement over how to store such dangerous wastes and political controversy over where to store them.

There are two reasons why we should reduce the amount of solid and hazardous wastes we produce. One reason is that at least 75% of these materials represent an unnecessary waste of resources. We can recycle up to 90% of the waste we produce, using current technology and waste recycling systems.

A second reason for reducing solid waste is that, in producing the products we discard, we create huge amounts of air pollution, greenhouse gases, water pollution, and land degradation.

A sustainable approach to solid waste is to reduce it, then reuse or recycle it, and finally safely dispose of what is left. We can deal with the solid wastes we create in two ways. One is waste management, in which we attempt to reduce the environmental impact of waste without seriously trying to reduce the amount of waste produced. It typically involves mixing wastes together and then transferring them from one part of the environment to another, usually by burying them, burning them, or shipping them to another location. The second approach is waste reduction, in which much less waste and pollution are produced, and the wastes that are produced are viewed as potential resources that can be reused, recycled, or composted.

There is no single solution to the solid waste problem. But we can cut solid wastes by reducing, reusing, and recycling This approach of waste reduction is based on three Rs:

-Reduce: consume less and live a simpler lifestyle.

-Reuse: rely more on items that can be used repeatedly instead of on throwaway items.

-Recycle: separate and recycle paper, glass, cans, plastics, metal, and other items, and buy products made from recycled materials.

Recycling is reprocessing discarded solid materials into new products. In addition to saving resources and reducing solid waste, recycling also reduces litter.

Households and workplaces produce five major types of materials that can be recycled: paper products, glass, aluminum, steel, and some plastics. These materials can be reprocessed in two ways. The first way is to recycle these materials into new products of the same type; turning used aluminum cans into new aluminum cans. In secondary way, waste materials are converted into different products. Used tires shredded and turned into rubberized road surfacing, newspapers reprocessed into cellulose insulation, and plastics reprocessed into various items, for example.

Just about anything is recyclable, but there are two key questions. First, are the items separated for recycling actually recycled? Sometimes they are mixed with other wastes and sent to landfills or incinerated. Second, will individuals complete the recycling loop by buying products that are made from recycled materials?

Three factors affect reuse and recycling. First, there is a faulty and misleading accounting system in which the market price of a product does not include the harmful environmental and health costs associated with the product during its life cycle. Second, there is an uneven economic playing field, because in most countries, resource-extracting industries receive more government tax breaks and subsidies than recycling and reuse industries get. Third, the demand and the price paid for recycled materials fluctuates, because buying goods made with recycled materials is not a priority for most governments, businesses, and individuals.

Advocates of recycling say that leveling the economic playing field is the best way to start encouragement for reuse and recycling programs. Governments can increase subsidies and tax breaks for reusing and recycling materials and decrease subsidies and tax breaks for making items from virgin resources.

Other strategies include greatly increasing the use of the fee-per-bag waste collection system and to encourage the government to purchase recycled products to help increase demand and lower prices. Governments can also pass laws requiring companies to take back and recycle or reuse packaging and electronic waste discarded by consumers, as is done in some countries already.

People can pressure the government to require labels on all products listing recycled content and the types and amounts of any hazardous materials they contain. This would help consumers to be better informed about the environmental consequences of buying certain products.

One reason for the popularity of recycling is that it helps to soothe people's consciences in a throwaway society. Many people think that recycling their newspapers and aluminum cans is all they need do to. Recycling is important, but reducing resource consumption and reusing resources are more effective ways to reduce the flow and waste of resources.

Shifting to a low-waste society requires individuals and businesses to reduce resource use and to recycle wastes at local, national, and global levels. In the United States, many individuals have organized to prevent the construction of hundreds of incinerators, landfills, treatment plants for hazardous and radioactive wastes, and polluting chemical plants in or near their communities. Health risks from incinerators and landfills, when averaged for the entire country, are quite low, but the risks for people living near such facilities are much higher.

Manufacturers and waste industry officials point out that something must be done with the toxic and hazardous wastes produced to provide people with certain goods and services. They contend that even if local citizens adopt a "not in my back yard" approach, the waste will always end up in someone's back yard. But many people don't accept this argument. They say the best way to deal with toxic and hazardous waste is to produce much less of it.

Environmental justice is an ideal whereby every person is entitled to protection from environmental hazards regardless of race, gender, age, national origin, income, social class, or any political factor. Studies have shown that too large a share of polluting factories, hazardous waste dumps, incinerators, and landfills in the United States are located in communities populated mostly by minorities and the working poor. These studies have also shown that, toxic waste sites in affluent communities have been cleaned up faster and more completely than sites in minority communities.

This environmental discrimination in the United States has led to a growing grassroots movement known as the environmental justice movement. This group has pressured the government, businesses, and environmental groups to become aware of environmental injustice and to act to prevent it. Even though progress has been made, there is still a long way to go.

Module 18 Goals

The following is a list of goals that should be met by the end of this module. The goals are very broad and, sometimes, general in nature. This was done to allow you to decide how much depth you'll want to go into for each goal.

Examining the Progress Check questions will also help focus your attention better on the goals.

1. Describe what solid waste and hazardous waste is, and why are they problems.
2. Explain how we should deal with solid waste.
3. Describe why reusing and recycling materials is so important.
4. Explain the advantages and disadvantages of burning or burying solid waste.
5. Explain how we should deal with hazardous waste.
6. Describe how we can make the transition to a more sustainable low-waste society.

Module 18 Key Words

Here's a list of key words for this module:

closed-loop recycling
hazardous waste
landfills
solid industrial waste
solid waste
waste management

Module 18 Progress Check

To check your understanding of the goals of this module, you should be able to answer the following questions.

1. What are the environmental problems associated with electronic waste?
2. What is the difference between solid waste, industrial solid waste, municipal solid waste, and hazardous (toxic) waste and give an example of each.
3. State two reasons for sharply reducing the amount of solid and hazardous waste we produce.
4. Describe the production of solid waste in the United States and what happens to such waste.
5. What are the differences between waste management, waste reduction, and integrated waste management?
6. Describe six ways in which industries and communities can reduce resource use, waste, and pollution.
7. Describe two approaches to recycling household solid wastes and evaluate each approach.
8. What are the major advantages and disadvantages of recycling?
9. What are three factors that discourage recycling?
10. Describe three ways to encourage recycling and reuse.
11. What are the major advantages and disadvantages of using incinerators to burn solid and hazardous waste?
12. What is the difference between open dumps and sanitary landfills?
13. What are the major advantages and disadvantages of burying solid waste in sanitary landfills?
14. Describe the regulation of hazardous waste in the United States under the *Resource Conservation and Recovery Act* and the *Comprehensive Environmental*

Response, Compensation, and Liability (Superfund) *Act.*

Module 19 Water Pollution

Module 19 Overview

Water pollution is any chemical, biological, or physical change in water quality that harms living organisms or makes water unusable. The chief sources of water pollution are agricultural activities, industrial facilities, and mining, but growth in population and resource use makes it increasingly worse.

Water pollution can come from single sources (point sources), or from larger, dispersed sources (non point sources). Point sources discharge pollutants at specific locations through drain pipes, ditches, or sewer lines into bodies of surface water. Since point sources are located at specific places, they are fairly easy to identify, monitor, and regulate. Most developed countries have laws that help to control point-source discharges of harmful chemicals into aquatic systems.

Nonpoint sources are broad, and diffuse areas from which pollutants enter surface water or air. Runoff of chemicals and sediments from cropland, livestock feedlots, logged forests, urban streets, parking lots, lawns, and golf courses are all non point sources of water pollution. Little progress in controlling water pollution from nonpoint sources has been because of the difficulty in identifying and controlling discharges from so many different sources.

The major leading causes of water pollution are agricultural activities, industry, and mining. Parking lots are a major source of nonpoint pollution for rivers and lakes because of grease, toxic metals, and sediments that collect on their impervious surfaces. Climate change will also contribute to water pollution in some areas. As climates change, some regions will get more precipitation and other areas will get less. Downpours will flush more harmful chemicals, plant nutrients, and microorganisms into waterways. Prolonged drought in other areas will reduce river flows that dilute wastes.

A major water pollution problem is exposure to infectious disease organisms from contaminated drinking water. More than 500 types of disease-causing bacteria, viruses, and parasites that can be transferred into water from the wastes of humans and animals have been identified.

Flowing rivers and streams can recover quickly from moderate levels of pollution through a combination of dilution and biodegradation of wastes by bacteria. This natural recovery process doesn't work when streams become overloaded with pollutants or when drought, damming, or water diversions reduce their flows.

Water pollution laws enacted in the 1970s have greatly increased the number and quality of wastewater treatment plants in the United States. These laws require industries to reduce or eliminate the point-source discharges of harmful chemicals into surface waters. This has enabled the United States to hold the line against increased pollution by pathogens and oxygen-demanding wastes in most of its streams. It's an impressive accomplishment given the country's increased economic activity, resource consumption, and population growth since passage of these laws.

The natural nutrient enrichment of a shallow lake, estuary, or slow-moving stream, mostly from runoff of plant nutrients such as nitrates and phosphates from surrounding land is called eutrophication. Cultural eutrophication occurs near urban or agricultural areas where human activities greatly accelerate the input of plant nutrients to a lake. According to the EPA, about one-third of the 100,000 medium to large lakes and 85% of the large lakes near major U.S. population centers have some degree of cultural eutrophication.

There are several ways to prevent or reduce cultural eutrophication. Expensive waste treatment to remove nitrates and phosphates before wastewater enters lakes can be used . We can banning or limit the use of phosphates in household detergents and other cleaning agents, and we can by employ soil conservation and land-use control to reduce nutrient runoff. As is the case, pollution prevention is more effective and usually cheaper in the long run than cleanup. A lake can usually recover from cultural eutrophication, though, if excessive inputs of plant nutrients are stopped.

Drinking water for about half of the U.S. population comes from groundwater and according to many scientists, groundwater pollution is a serious threat to health. Common pollutants like fertilizers, pesticides, gasoline, and organic solvents have seeped into the groundwater from many sources.

Once a pollutant contaminates groundwater, it fills the aquifer's porous layers of sand, gravel, or bedrock like water saturates a sponge. This makes removal of the contaminant difficult and costly. The slowly flowing groundwater disperses the pollutant in a wide are of contaminated water that eventually reaches a well and is pumped out to be used as drinking water or water used to irrigate crops.

Contaminated groundwater cannot cleanse itself of degradable wastes as quickly as flowing surface water can. Groundwater flows slowly and contaminants are not diluted and dispersed effectively. Groundwater also has much lower concentrations of dissolved oxygen and smaller populations of decomposing bacteria. As a result, it can take decades for contaminated groundwater to cleanse itself of slowly degradable wastes.

The great majority of ocean pollution originates on land and includes oil and other toxic chemicals and solid wastes, which threaten aquatic species and other wildlife and disrupt marine ecosystems. Coastal areas bear the brunt of our enormous inputs of pollutants and wastes into the ocean, which is not surprising, because about 40% of the world's population lives on or near the coast.

In deeper waters, the oceans can dilute, disperse, and degrade large amounts of raw sewage and other types of degradable pollutants. Some advocate the dumping of sewage sludge and other harmful wastes into the deep ocean than to bury them. Since we know less about the deep ocean than we do about the moon, opponents of this idea disagree.

There is also the problem of cruise ship pollution. A cruise ship can 2,000 or more passengers and 1,000 crew members. They can generate as much waste as a small city. Much of this waste is highly toxic. They also generate huge amounts of plastic garbage and waste oil. For decades, cruise ships and other ocean vessels, have been dumping their wastes at sea.

In U.S. waters, such dumping is illegal, even though some ships dump secretively at night. Since 2002, companies that have been caught in the act of illegal dumping have been fined millions of dollars so ship owners have reduced this illegal activity. Crude and refined petroleum reach the ocean from a number of sources and become highly disruptive pollutants.

Perhaps half of the oil reaching the oceans is waste oil; dumped, spilled, or leaked onto the land or into sewers.

The volatile hydrocarbons in oil immediately kill aquatic organisms, especially in their larval stage. The other chemicals in oil form tar-like globs that float on the surface and coat the feathers of birds and the fur of marine mammals. The coating destroys their natural heat insulation and buoyancy, causing many of them to drown or die of exposure from loss of body heat.

The heavy oil components sink to the ocean floor or wash into estuaries smother bottom-dwelling organisms such as crabs, oysters, mussels, and clams, or make them unfit for human consumption. Some oil spills have killed coral reefs. Populations of marine life can recover from exposure to large amounts of crude oil within about 3 years. But recovery from exposure to refined oil can take anywhere from 10–20 years, especially in estuaries and salt marshes.

While working with nature to treat sewage, cutting resource use and waste, reducing poverty, and slowing population growth are efforts aimed at reducing water pollution, laws can help by preventing it. The Federal Water Pollution Control Act of 1972 (re-named the Clean Water Act 1977) and the 1987 Water Quality Act are the basis of U.S. efforts to control water pollution. The Clean Water Act sets standards for levels of key water pollutants and requires polluters to get permits limiting how much of various pollutants they can discharge into aquatic systems.

As a result of this legislation, between 1992 and 2002 the number of Americans served by community water systems that met federal health standards increased from 79% to 94%, the percentage of U.S. stream lengths found to be fishable and swimmable increased from 36% to 60% of those tested, the amount of topsoil lost through agricultural runoff was cut by about 1 billion tons annually, the proportion of the U.S. population served by sewage treatment plants increased from 32% to 74%, and annual wetland losses decreased by 80%.

This illustrates that legislation along with increased public awareness of the problems of water pollution, can control the quality of water resources.

Module 19 Goals

The following is a list of goals that should be met by the end of this module. The goals are very broad and, sometimes, general in nature. This was done to allow you to decide how much depth you'll want to go into for each goal.

Examining the Progress Check questions will also help focus your attention better on the goals.

1. Identify the causes and effects of water pollution.
2. Explain the major water pollution problems in streams and lakes.
3. Identify the major pollution problems affecting groundwater and other drinking water sources.
4. Describe the major water pollution problems affecting oceans.
5. Explain how we can best deal with water pollution.

Module 19 Key Words

Here's a list of key words for this module:

non-point sources
point sources
primary sewage treatment
secondary sewage treatment
water pollution

Module 19 Progress Check

To check your understanding of the goals of this module, you should be able to answer the following questions.

1. What is the difference between point sources and non-point sources of water pollution and give an example of each.
2. Identify nine major types of water pollutants and give an example of each.
3. Describe the chemical and biological methods that scientists use to measure water quality.
4. Describe the difference between eutrophication and cultural eutrophication.
5. List three ways to prevent or reduce cultural eutrophication.
6. What are the major sources of groundwater contamination in the United States?
7. Identify and describe the U.S. laws for protecting drinking water quality.
8. What are the environmental problems caused by the widespread use of bottled water?
9. How serious is oil pollution of the oceans, what are its effects, and what can be done to reduce such pollution.
10. Describe how primary sewage treatment and secondary sewage treatment are used to help purify water.
11. List six ways to prevent water pollution.
12. List five steps that can be taken to reduce water pollution.

Module 20 Air Pollution

Module 20 Overview

Air pollution is the presence of chemicals in the atmosphere in concentrations high enough to harm organisms, ecosystems, or human-made materials.

Air pollutants come from both natural and human sources. Natural sources would be dust blown by wind, pollutants from wildfires and volcanic eruptions, and volatile organic chemicals released by some plants. Human sources of air pollutants occur in industrialized areas where people, cars, and factories are concentrated. Most of these pollutants come from the burning of fossil fuels. As it turns out, air pollution the earth's oldest environmental problems.

Prior to the widespread use of coal as a source of power, most air pollution problems centered on the burning of wood. During the Middle Ages most urban areas were clouded with a haze of wood smoke.

With the Industrial Revolution coal became the single most important source of energy. It was plentiful, cheap, and relatively easy to mine. As a result of sung coal in industry and in homes, many urban areas became so polluted that respiratory diseases like asthma and bronchitis dramatically increased. Among the European cities, London was especially prone to severe air pollution problems. By the 1850s the smoke and fog in London would engulf the city for weeks on end. By the early 1900s the word smog was used widely to describe the air pollution in the city.

In 1880 a smog episode killed 2,200 people. In 1911 another 1,100 lost their lives. December of 1952 saw the worst episode of smog killing between 4,000 and 11,000 people over a five day period. The British Parliament passed the Clean Air Act of 1956 and before the effects of the new law could be materialized, air pollution disasters occurred again in 1956, 1957, and 1962 killing additional hundreds of people.

The United States was not immune to its own air pollution problems. The large coal deposits in Appalachia made our own Industrial Revolution possible. By the 1940s all large urban areas suffered from some air pollution. But industrial cities like Pittsburgh were worse. The first air pollution disaster occurred in a small town south of Pittsburgh were pollutants from the surrounding coal-burning factories got trapped in a dense fog that lasted for 5 days. In that period of time 6,000 residents became sick and 20 died. By the late 1960s hazy skies were common over many U.S. Cities and industrial areas. It was at that time that the government began to take steps to reduce the pollution through legislation.

Air pollution consists of a number of components. First are the carbon oxides. Compounds lie carbon monoxide and carbon dioxide fall into this category. Both are generated from the burning of fossil fuels and many are aware of the dangers of carbon monoxide poisoning.

Next are the nitrogen oxides. Nitric oxide is formed in car engines and when it is released it combines with oxygen to form nitrogen dioxide. Nitrogen dioxide is a reddish-brown gas. Sometimes nitrogen dioxide will combine with water to form nitric acid.

Then there is sulfur dioxide and sulfuric acid. Sulfur dioxide is a colorless gas with an irritating odor. About two-thirds of the sulfur dioxide comes from human sources, mostly from the burning of sulfur-containing coal in electric power and industrial plants. Sulfur dioxide, sulfuric acid droplets, and sulfate particles reduce visibility and aggravate breathing problems. Sulfur dioxide and sulfuric acid damage crops, trees, soils, and aquatic life in lakes. They also corrode metals, damage paint, paper, leather, and stone on buildings and statues.

Finally there are the particulates. This consists of a variety of solid particles and liquid droplets small and light enough to remain suspended in the air for long periods. About 62% of the particulates in outdoor air comes from natural sources such as dust, wild fires, and sea salt. The remaining 38% comes from human sources such as coal-burning power and industrial plants, motor vehicles, plowed fields, road construction, unpaved roads, and tobacco smoke.

According to the American Lung Association, more than 2,000 studies published since 1990 link particulates with adverse health effects. Particulates have been shown to irritate the nose and throat, damage the lungs, aggravate asthma and bronchitis, and shorten life. Particulates also reduce visibility, corrode metals, and discolor clothes and paints.

Of increasing concern is indoor air pollution. We normally don't consider the air in homes to be polluted but since 1990 the EPA has placed indoor air pollution at the top of the 18 sources of cancer risk. The risk factors of indoor air pollution include not only smoke from cigarettes, but pesticides used in the home, chemicals from household solvents, mold and fungal spores, and formaldehyde which is found in many building materials. Long term exposure to these factors results in severe respiratory problems or other health problems as a result of the immune system becoming weaker because of exposure.

The U.S. Congress passed three Clean Air Acts in 1970, 1977, and 1990. These laws allow the federal government to established air pollution regulations that focus on key pollutants. The EPA established national ambient air quality standards for six outdoor pollutants; carbon monoxide, nitrogen oxides, sulfur dioxide, particulates, ozone, and lead. One limit is set to protect human health and another limit is intended to prevent environmental and property damage.

The EPA also established national emission standards for more than 188 hazardous air pollutants that may cause serious health and ecological effects. An important public source of information about hazardous air pollutants is the annual Toxics Release Inventory (TRI). The TRI law requires 21,500 refineries, power plants, hard rock mines, chemical manufacturers, and factories to report their releases and waste management methods for 667 toxic chemicals.

The TRI, which is now available on-line, lists this information by community (the web address is listed in Web Resources). This allows individuals and local groups to evaluate potential threats to their health. Reported emissions of toxic chemicals have dropped sharply since the first TRI report was released in 1988.

Module 20 Goals

The following is a list of goals that should be met by the end of this module. The goals are very broad and, sometimes, general in nature. This was done to allow you to decide how much depth you'll want to go into for each goal.

Examining the Progress Check questions will also help focus your attention better on the goals.

1. Identify the structure of the atmosphere.
2. Describe the major outdoor air pollution problems.
3. Explain why acid deposition and why it's a problem.
4. Describe the major indoor air pollution problems.
5. Describe the ways we should deal with air pollution.

Module 20 Key Words

Here's a list of key words for this module:

acid disposition
air pollution
atmospheric pressure
carbon oxides
ozone layer
particulates
primary pollutants
secondary pollutants
smog
stratosphere
temperature inversion
troposphere

Module 20 Progress Check

To check your understanding of the goals of this module, you should be able to answer the following questions.

1. Define density, atmospheric pressure, troposphere, stratosphere, and ozone layer.
2. Describe how the troposphere and stratosphere differ.
3. What is air pollution?
4. What is the difference between primary pollutants and secondary pollutants and give an example of each.
5. List the major outdoor air pollutants and their harmful effects.
6. List and describe five natural factors that help to reduce outdoor air pollution and six natural factors that help to worsen it.
7. What are the top four indoor air pollutants in the United States?
8. Describe and summarize the air pollution laws in the United States.
9. Summarize the major ways to reduce emissions from power plants and motor vehicles.
10. Discuss the relationship between the Asian Brown Cloud and the ways in which people have violated the four scientific principles of sustainability.

Module 21 Climate Change

Module 21 Overview

The one environmental problem that gets a lot of media attention these days is climate change. It is the one thing that can have a significant impact on all societies around the globe. It is also one of those environmental factors that once it reaches a tipping point, will have a cascading effect, like a stack of dominoes, and no one will be immune.

Climate change is neither new nor unusual. Over the past 4.5 billion years, the earth's climate has been altered by volcanoes, changes in solar input, continents slowly moving, and impacts by large meteors and asteroids. Over the past 900,000 years, the earth has gone through long periods of global cooling and global warming. These alternating cycles of freezing and thawing are known as glacial and interglacial periods.

For roughly 10,000 years, we have been in an interglacial period with fairly stable climates and a fairly steady average global surface temperature. These conditions have allowed agriculture to flourish, cities to grow, and the human population to increase. But all of this seems to be changing.

In order to understand what climate change is, you need to first understand the natural greenhouse effect.

Solar energy warms the earth's lower atmosphere and surface. This happens when some of the solar energy that is absorbed by the earth radiates into the atmosphere as infrared radiation (heat). About 1% of the earth's lower atmosphere is composed of the greenhouse gases water vapor, carbon dioxide, methane, and nitrous oxide. As the earth radiates heat into the atmosphere it causes molecules of these greenhouse gases to vibrate and release more heat into the lower atmosphere and warms the lower atmosphere and the earth's surface, which in turn then affects the earth's climate.

The Swedish chemist Svante Arrhenius was the first scientist to recognize this natural greenhouse effect in 1896. Since then, many tests and experiments have confirmed this effect, which is now one of the most widely accepted theories in the study of the atmosphere.

As a result of the Industrial Revolution, we have seen increases in the concentration of the greenhouse gasses carbon dioxide, methane, and nitrous oxide in the lower atmosphere. These gasses come from agriculture, deforestation, and burning of fossil fuels. By measuring levels of carbon dioxide and methane in bubbles at various depths in ancient glacial ice scientists observed that changes in the levels of these gasses in the lower atmosphere correlate fairly closely with changes in the average global temperature near the earth's surface during the past 400,000 years, and with changes in the global sea level.

Today carbon dioxide levels are increasing at an exponential rate. The concentration of carbon dioxide emitted by the burning of fossil fuels has risen from a level of 280 parts per million since the beginning of the Industrial Revolution 275 years ago, to 384 parts per million in 2007. The levels continue to increase about 2 parts per million each year.

The concern is that if these levels continue to increase at the current rate of 3.3% a year, carbon dioxide levels will rise to about 560 ppm by 2050 and to 1,390 ppm by 2100. By all accounts this will bring significant changes to the earth's climate and cause major ecological and economic disruption. According to studies that have been done we should try to prevent carbon dioxide levels from exceeding 450 ppm. Many scientists agree that this is an irreversible tipping point that could set into motion large-scale climate changes for hundreds to thousands of years. Some scientists even advocate that we need to reduce concentrations of carbon dioxide to 350 parts per million, which is below current levels, to help stabilize the earth's climate.

Many people do not realize how much of a significant role the oceans play in climate change. The oceans act as a sponge when it comes to absorbing carbon dioxide. In fact the oceans have already absorbed about half of all the carbon dioxide released to the atmosphere since the beginning of the Industrial Revolution. Today, the oceans remove about 25–30% of the carbon dioxide released into the lower atmosphere. This helps stabilize global temperatures and moderates climate patterns.

The problem is that the solubility of carbon dioxide in ocean water decreases with increasing temperature. So as the oceans heat up, some of the dissolved carbon dioxide can be released into the lower atmosphere. This can speed up global warming and climate a positive feedback loop. Global measurements have shown that the upper portion of the ocean warmed by 0.6–1.2F° during the last century.

One of the big unknowns in climate change is the effect of cloud cover. Warmer temperatures will increase evaporation of surface water and create more clouds. Depending on their content and reflectivity, these additional clouds could have one of two effects. First, an increase in thick, continuous light-colored clouds at low altitudes can decrease surface warming by reflecting more sunlight back into space. Secondly, an increase in thin, discontinuous clouds at high altitudes could warm the lower atmosphere by trapping more heat.

No one knows for sure what the effect of climate change will be, but there are a number of scenarios that indicate what those changes might be.

Over the past few decades there has been a significant increase in drought areas. If this trend continues, there will be less moisture in the soil; stream flows and surface water will decline; growth of trees and other plants will slow, which will reduce carbon dioxide removal from the atmosphere and intensify global warming; forest and grassland fires will increase, which will add more carbon dioxide to the atmosphere; water tables will fall with more evaporation and irrigation; and some lakes and seas will shrink or disappear. Dry climate biomes, such as savannas, chaparral, and deserts, will increase. The effects of prolonged drought over several decades create conditions that accelerate global warming and climate change and lead to even more drought.

Climate models predict that global warming will be the most severe in the polar regions. Ice and snow in the polar regions help to cool the earth by reflecting incoming solar energy. The melting of the ice and snow exposes much darker land and sea, which absorb more solar energy. This will cause polar regions to warm faster than lower latitudes, which will accelerate global warming. The resulting climate change will melt more sea ice, which will raise atmospheric temperatures more, and faster, in a runaway positive feedback loop.

Loss of arctic sea ice reduces long-term average rainfall and snowfall in areas like the already dry American West, which will affect food production by reducing the availability of irrigation water. A loss of arctic sea ice could also increase long-term average precipitation and flooding in western and southern Europe.

The media loves to pay up rising sea levels as a result of global warming. Stories about the major cities along the eastern seaboard completely underwater seem to run on a regular basis. But the underlying fact is that sea levels are indeed rising. How much and how long seems to be still open to debate. A study done in 2007 said that sea levels will rise anywhere from 0.6-1.9 feet by the end of this century. Another study in 2008 said that sea levels will rise 3.3-6.6 feet before the end of the century. An increase in sea levels of 1.6 feet will affect over 150 million people and cause over $35 trillion dollars in damage. Even though sea levels will rise slowly, the long term impact cannot be ignored before it's too late.

As global temperatures rise, the frozen tundra will melt. This will release all the trapped carbon dioxide and methane stored in the frozen earth, further increasing the levels of those greenhouse gasses.

Because the world's glaciers are melting as a rapid rate, many scientists fear that the introduction of all the freshwater could very well change ocean currents. The impact this will have on global climates is really unknown.

Global climate changes can cause in increase in severe weather. It's possible that we'll see an increase in the incidence of extreme weather such as heat waves and droughts in some areas, which could kill large numbers of people, reduce crop production, and expand deserts. And because a warmer atmosphere can hold more moisture, other areas will experience increased flooding from heavy and prolonged precipitation.

Of course all of this has an impact on biodiversity. As global climates change, so do biomes and ecosystems. Organisms will adapt to new environments and some will become extinct.

There's no question that dealing with climate change is going to be difficult. It's becoming more and more clear that addressing climate change is going to be one of the most urgent scientific, political, economic, and ethical issues that humanity faces.

Module 21 Goals

The following is a list of goals that should be met by the end of this module. The goals are very broad and, sometimes, general in nature. This was done to allow you to decide how much depth you'll want to go into for each goal.

Examining the Progress Check questions will also help focus your attention better on the goals.

1. Explain how the earth's temperature and climate will change in the future.
2. Identify some possible effects of a warmer atmosphere.
3. Identify what can be done to slow climate change.
4. Explain how ozone became depleted in the stratosphere and what can done about it.

Module 21 Key Words

The key words for this module you've already seen so there's none for this module. Hooray!!

Module 21 Progress Check

To check your understanding of the goals of this module, you should be able to answer the following questions.

1. Describe global warming and cooling patterns over the past 900,000 years and during the last century.
2. How do scientists get information about past temperatures and climates?
3. What is the greenhouse effect and why is it so important to life on the earth?
4. What is the role played by oceans in the regulation of atmospheric temperatures?
5. Describe how each of the following might affect global warming and its resulting effects on global climate: (a) cloud cover and (b) air pollution.
6. What are four major strategies for slowing projected climate change?
7. What are the pros and cons of the Kyoto Protocol?
8. Describe how human activities have depleted ozone in the stratosphere, and list five harmful effects of such depletion.
9. What has the world done to help reduce the threat from ozone depletion in the stratosphere?
10. Describe how the four scientific principles of sustainability can be applied to deal with the problems of climate change and ozone depletion.

Unit 6 - Environmental Sustainability

Module 22 Environment and Cities

Module 22 Overview

We know that half of the world's population lives in urban areas. That's over 3.5 billion people and growing. The problems that big cities face will have an everlasting impact on global ecosystems. Urbanization is defined as the creation and growth of cities and their surrounding developed land. Urban growth is the rate of increase of urban populations. Today 79% of Americans and almost 50% of the world's people live in urban areas.

Urban areas grow in two ways; by natural increase and by immigration, mostly from rural areas. Rural people are pulled to urban areas in search of jobs, food, housing, educational opportunities, better health care, entertainment, and freedom from religious, racial, and political conflicts. Some are also pushed from rural to urban areas by factors such as poverty, lack of land for growing food, declining agricultural jobs, famine, and war. People are also pushed and pulled to cities by government policies that favor urban over rural areas.

There are four major trends in urban population dynamics. First, the proportion of the world's population living in urban areas is increasing. Between 1850 and 2008, the percentage of people living in urban areas increased from 2% to almost 50%. Much of the future growth will occur in already overcrowded and stressed cities in developing countries.

Second, urban areas are expanding rapidly in number and size. Each week 1 million people are added to the world's urban areas. By 2015, the number of urban areas with a million or more people is projected to increase from 400 to 564. There are also 18 megacities, cities with 10 million or more people. Fifteen of them are in developing countries. Megacities will soon be eclipsed by hypercities with more than 20 million people each. Tokyo, with 26.5 million people, is the only city in this category.

Third, urban growth is much slower in developed countries than in developing countries. Fourth, poverty is becoming increasingly urbanized. It's estimated that at least 1 billion people in developing countries live in crowded and unsanitary slums and shantytowns. Within 30 years this number may double.

Urbanization does have some benefits from an economic standpoint. Cities are the centers of economic development, innovation, education, technological advances, and jobs. They serve as centers of industry, commerce, and transportation. Urban residents tend to live longer than do rural residents and have lower infant mortality rates and fertility rates. They have better access to medical care, family planning, education, and social services than do their rural counterparts. However, the health benefits of urban living are usually greater for the rich than for the poor.

Urban areas also have some environmental advantages. Recycling is more economically feasible because concentrations of recyclable materials and funding for recycling programs tend to be higher in urban areas. Concentrating people in cities helps to preserve biodiversity by reducing the stress on wildlife habitats. And central cities can save energy if residents rely more on energy-efficient mass transportation, walking, and bicycling.

But the disadvantages of urbanization are significant and cannot be overlooked. Each year, about 63 million people are added to urban areas. These intense population pressures make most of the world's cities more environmentally unsustainable every year. Even in more sustainable cities this pressure is taking its toll.

Urban populations occupy about 2% of the earth's land area and consume 75% of its resources. Because of the high resource input of food, water, and materials and high waste output, most of the world's cities have huge ecological footprints and are not self-sustaining.

Most cities do not benefit from vegetation that would absorb air pollutants, give off oxygen, cool the air through transpiration, provide shade, muffle noise, provide wildlife habitats, and give aesthetic pleasure. As cities grow and water demands increase, expensive reservoirs and canals must be built and deeper wells must be drilled. This deprives rural areas of surface water and depletes groundwater.

Because of high population densities and high resource consumption, cities produce most of the world's air pollution, water pollution, and solid and hazardous wastes. Pollutant levels are higher because pollution is produced in a smaller area and cannot be dispersed and diluted as readily as pollution produced in rural areas.

Most urban dwellers are subjected to noise pollution, which is any unwanted, disturbing, or harmful sound that impairs or interferes with hearing. Noise pollution can cause stress, hamper concentration and work efficiency, or causes accidents. Cities generally are warmer, rainier, foggier, and cloudier than suburbs. The enormous amount of heat generated by cars, factories, furnaces, lights, air conditioners, and heat-absorbing dark roofs and streets in cities creates an urban heat island that is surrounded by cooler suburban and rural areas.

In recent years, builders have begun using a pattern known as cluster development. In cluster development, high-density housing units are concentrated on one portion of a parcel and the rest of the land is used for shared open space. Done properly, residents live with more open and recreational space, aesthetically pleasing surroundings, and lower heating and cooling costs because some walls are shared. Developers can also cut their costs for site preparation, roads, utilities, and other forms of infrastructure.

Some developers are using an approach called new urbanism. This is a modern form of what could be called old villageism. New urbanism develops entire villages and promotes mixed-use neighborhoods within existing cities. There are many examples of this approach across the United States.

So how can cities become more sustainable and livable? One approach is called an ecocity. An ecocity is one where people choose walking, biking, or mass transit for most transportation needs. They recycle or reuse most of their wastes. They grow much of their food and protect biodiversity by preserving surrounding land.

Examples of cities that have attempted to become more environmentally sustainable and to become more livable include Portland, Oregon; Davis, California; Olympia, Washington; Chattanooga, Tennessee; Waitakere City, New Zealand; Stockholm, Sweden; Helsinki, Finland; and Leichester, England.

Module 22 Goals

The following is a list of goals that should be met by the end of this module. The goals are very broad and, sometimes, general in nature. This was done to allow you to decide how much depth you'll want to go into for each goal.

Examining the Progress Check questions will also help focus your attention better on the goals.

1. Identify the major population trends in urban areas.
2. Discuss the major urban resource and environmental problems.
3. Explain how transportation affects urban environmental impacts.
4. Explain how important urban land use planning is.
5. Discuss how cities can become more sustainable and livable.

Module 22 Key Words

Here's a list of key words for this module:

land-use planning
noise pollution
urban growth
urban sprawl

Module 22 Progress Check

To check your understanding of the goals of this module, you should be able to answer the following questions.

1. What is the difference between urbanization and urban growth?
2. Describe two factors that increase the population of a city.
3. List four trends in global urban growth.
4. Describe four phases of urban growth in the United States.
5. What is urban sprawl?
6. List six factors that have promoted urban sprawl in the United States.
7. List five undesirable effects of urban sprawl.
8. Explain why most cities and urban areas are not sustainable.
9. List four ways to reduce dependence on motor vehicles.
10. Describe the major advantages and disadvantages of relying more on (a) bicycles, (b) mass transit rail systems, and (c) bus rapid transit systems within urban areas, and (d) rapid-rail systems between urban areas.
11. What are the five guiding principles of new urbanization?

Module 23 Economics of Sustainability

Module 23 Overview

If the goal is to attain, and maintain, a sustainable environment, then you have to look at what it costs. This is a very complex issue for many governments, both national and local, as they try to find the funds for paying for the programs. There are many environmental economic models that appear to be contradictions to the traditional economic models we are used to. Here's an example of one approach.

Environmental economists argue that an environmentally honest market system would include the harmful environmental and health costs of goods and services in their market prices to reflect as close as possible their full costs. In other words, the good we buy today are not priced honestly because e they do not include the environmental cost of producing that product.

According to this approach, full-cost pricing would reduce resource waste, pollution, and environmental degradation and improve human health by encouraging producers to invent more resource-efficient and less-polluting methods of production. Jobs and profits would be lost in environmentally harmful businesses as consumers chose environmentally friendly products, but jobs and profits would be created in environmentally beneficial businesses. Such shifts in profits and in types of businesses and jobs is the way market-based capitalism is supposed to work.

Why isn't full-cost pricing used more widely? First, many producers of harmful and wasteful products would have to charge more, and some would go out of business. Obviously they will oppose such pricing. Second, it's difficult to estimate many environmental and health costs. Third, because these harmful costs are not included in the market prices of goods and services, most consumers don't connect them with the things they buy. Correcting this will require consumer education. Few companies will volunteer to reduce their short-term profits to become more environmentally responsible unless it eventually becomes more profitable for them.

Product eco-labeling and certification programs already in place have encouraged companies to develop green products and services. This helps consumers select environmentally beneficial products and services. Eco-labeling programs have been developed in Europe, Japan, Canada, and the United States. The U.S. Green Seal labeling program has certified more than 300 products as environmentally friendly.

One way to encourage the shift to full-cost pricing is to phase out environmentally harmful subsidies and tax breaks. These subsidies and tax breaks cost the world's taxpayers at least $2 trillion a year. These subsidies distort the economic playing field and create a huge economic incentive for unsustainable resource waste, depletion, and degradation.

On paper, phasing out subsidies seems like a great idea. In reality, though the economically and politically powerful interests receiving them want to keep, and if increase, these hidden economic benefits. In addition to lobbying to keep or increase their subsidies, such businesses often lobby against subsidies and tax breaks for more environmentally beneficial competitors. Environmentally destructive subsidies and tax breaks will continue until enough people work together to counteract such lobbying by electing, and then influencing, elected officials who support phasing out environmentally harmful subsidies.

Making the shift to more sustainable economies will require governments and industries to greatly increase their spending on research and development. Analysts also urge major corporations to play a much greater role in shifting to an eco-economy by becoming more socially and ecologically responsible, for both ethical and economic reasons. General Electric is the world's ninth largest corporation. In 2004 it launched its "ecoimagination plan," which commits the company to greatly increasing its investments in green technology research.

Educational institutions around the world can play a key role in making the transition to more sustainable economies. They can do this by giving all students a basic environmental education and by developing business schools that integrate sustainable business planning and management into their curricula.

In the United States, there are business schools dedicated to teaching sustainable business. These business schools emphasize business and environmental sustainability and social responsibility. Graduates from these business schools talk about the triple bottom line: people, planet, and profit.

Module 23 Goals

The following is a list of goals that should be met by the end of this module. The goals are very broad and, sometimes, general in nature. This was done to allow you to decide how much depth you'll want to go into for each goal.

Examining the Progress Check questions will also help focus your attention better on the goals.

1. Explain how economic systems are related to the biosphere.
2. Explain how we put values on natural capital, pollution control, and resource use.
3. Identify how we use economic tools to deal with environmental problems.
4. Discuss how reducing poverty helps us to deal with environmental problems.
5. Explain how we can make the transition to more environmentally sustainable economies.

Module 23 Key Words

Here's a list of key words for this module:

high-throughput economies
human resources
low-throughput economies
manufactured resources
natural capital

Module 23 Progress Check

To check your understanding of the goals of this module, you should be able to answer the following questions.

1. What is an economic system?
2. Why can a sustainability revolution be also an economic revolution?
3. What are the differences between natural capital, human capital, and manufactured capital?
4. Compare how neoclassical economists and ecological and environmental economists view economic systems.
5. List eight strategies that ecological and economic economists would use to make the transition to more sustainable eco-economies.
6. What would the benefits of shifting from environmentally unsustainable to more environmentally sustainable government subsidies and tax breaks be?
7. How is poverty related to population growth and environmental degradation?
8. What are the environmental benefits of selling services instead of goods?
9. Give two examples of the above approach.
10. List six ways to shift to more environmentally sustainable economies.

Module 24 Politics of Sustainability

Module 24 Overview

Businesses and industries are supposed to thrive on change and innovations that lead to new technologies, products, and opportunities for profits. This process is free enterprise and can lead to higher living standards for many people, but it can also create harmful impacts on other people and on the environment.

Government on the other hand, especially democratic governments are supposed to act as a brake on business enterprises that might result in harm to people or the environment. Achieving the right balance between free enterprise and government regulation is not easy. Too much government intervention can strangle business enterprise and innovation. Too little can lead to environmental degradation and social injustices and even to a takeover of the government by business and global trade interests.

In today's global economy it's not uncommon to find multinational corporations that have budgets larger than the budgets of many countries. These multinational corporations have great economic and political power over national, state and local governments, and ordinary citizens.

In most constitutional democracies, political institutions are designed to allow for gradual change to ensure economic and political stability. In the United States rapid and destabilizing change is curbed by a system of checks and balances that distributes power among the three branches of government.

Elected and appointed government officials must deal with pressure from many competing special-interest groups. Each group advocates passing laws, providing subsidies or tax breaks, or establishing regulations favorable to its cause. They also bring pressure to weaken or repeal laws, subsidies, tax breaks, and regulations unfavorable to its position.

Several features of democratic governments interfere with the ability to deal with environmental problems. Problems such as climate change and biodiversity loss, for example, are complex and difficult to understand. These problems have long-lasting effects, are interrelated, and require integrated, long-term solutions that emphasize prevention. Because elections are held every few years, politicians seeking reelection focus on short-term, isolated issues rather than on complex, time-consuming, and long-term problems.

As we have experienced in the United States, developing an environmental policy in a controversial process. Passing a law does not make a policy. Funds need to be appropriated and then a government agency needs to draw up regulations and ways to implement those regulations. Throughout the entire process there is the possibility for special-interest groups to exert their influence so that what was originally proposed is either watered-down or very different.

The courts, of course, enforce the violations of laws. But environmental lawsuits are difficult to win. Plaintiffs bringing the suit must establish that they have the legal right to do so in a particular court. To have that right, plaintiffs must show that they have personally suffered health problems or financial losses allegedly caused by the defendant's actions. Bringing these kinds of lawsuits to court are usually too expensive for most individuals.

Public interest law firms cannot recover attorneys' fees unless Congress has specifically authorized it. In contrast, corporations can reduce their taxes by deducting their legal expenses. The legal playing field has become uneven and, in financial terms, is stacked against individuals and groups of private citizens filing environmental lawsuits.

To collect damages resulting from a nuisance or an act of negligence, plaintiffs must establish that they have been harmed in some significant way and that the defendant caused the harm. Proving this can be difficult and costly. Suppose a company is sued for causing cancer in individuals who get their drinking water from a river into which the company dumped hazardous chemicals. If other industries and cities also dump waste into that river, establishing that the company is the culprit requires expensive investigation, scientific research, and expert testimony. In addition, it's hard to establish that a particular chemical caused the plaintiffs' cancers.

Many states have statutes of limitations, which limit how long a plaintiff can take to sue after a particular event occurs. These statutes make it virtually impossible for victims of cancer, which may take 10–20 years to develop, to file a negligence suit. The court, or series of courts if the case is appealed, can take years to reach a decision. Unless the court issues a temporary injunction against it until the case is decided the defendant may continue the allegedly damaging action.

Sometimes corporations and developers file what are mown as SLAPPs (strategic lawsuits against public participation) against citizens who publicly criticize a business for some activity. There was such a case in Texas when a woman publicly called a nearby landfill "a dump" and the landfill owners sued her husband for $5 million for failing to "control his wife." Most SLAPPs are not meant to be won, but are intended to intimidate individuals and activist groups.

To counteract SLAPPs, some have fought back with countersuits and have been awarded damages. A Missouri woman was sued for criticizing a medical waste incinerator and won an $86 million judgment against the incinerator's owner. But even after paying these kinds of awards, corporations generally save money because they can write off legal and liability insurance costs as business expenses.

U.S. environmental laws have been highly effective, especially in controlling pollution. Since 1980, though, a well-organized and well-funded movement is trying to weaken or repeal existing environmental laws and to change the ways in which public lands are used.

Three major groups are strongly opposed to various environmental laws and regulations: some corporate leaders who view the laws as threats to their profits, wealth, and power; individual citizens who see them as threats to their property rights and jobs; and state and local governments who resent having to implement federal laws and regulations with little or no federal funding.

Since 2000, efforts to weaken environmental laws and regulations have escalated. Most U.S. environmental laws and regulatory agencies have been weakened by a combination of executive orders and congressional actions. Many regulatory agencies have been staffed with officials who favor weakening them, decreasing their funding, ignoring sound scientific consensus, and stifling dissent. Environmental leaders warn that the entire environmental legal and regulatory structure, built with bipartisan consensus between 1965 and 1980 is being systematically undermined.

Most Americans are unaware that the increasingly weakened foundation of the nation's environmental laws and regulations are in jeopardy. Polls show that more than 80% of the U.S. public strongly support environmental laws and regulations and don't want them weakened. Polling data also show that the percentage of Americans who say they worry about the environment "a great deal" or "a fair amount" has increased from 62% to 77% between 2004 and 2006. But polls also show that less than 10% of the U.S. public consider the environment to be one of the nation's most pressing problems. As a result, environmental concerns often do not get transferred to the ballot box or the pocketbook.

The spearhead of the global conservation, environmental, and environmental justice movements are the tens of thousands of nonprofit, nongovernmental organizations working at the international, national, state, and local levels. The growing influence of these organizations is the most important trends influencing environmental decisions and policies.

Module 24 Goals

The following is a list of goals that should be met by the end of this module. The goals are very broad and, sometimes, general in nature. This was done to allow you to decide how much depth you'll want to go into for each goal.

Examining the Progress Check questions will also help focus your attention better on the goals.

1. Discuss the role government plays in making the transition to more sustainable societies.
2. Explain how environmental policy is made.
3. Explain the role environmental law plays in dealing with environmental problems.
4. Identify the role of major environmental groups.
5. Discuss how we can implement more sustainable and just environmental policies.

Module 24 Key Words

Here's a list of key words for this module:

administrative law
arbitration
common law
environmental law
environmental policies
lobbying
mediation
statutory law

Module 24 Progress Check

To check your understanding of the goals of this module, you should be able to answer the following questions.

1. Describe two features of democratic governments that hinder their ability to deal with environmental problems.
2. Identify and describe three major U.S. environmental laws.
3. What are four major types of public lands in the United States and describe the controversy over the management of these lands.
4. What does it mean to say that we should think globally and act locally? Give an example of such an action.
5. Explain the differences between a statutory law, an administrative law, and a common law.
6. List six reasons why it is difficult to win an environmental lawsuit.
7. List three general types of U.S. environmental laws.
8. What is an environmental impact statement?
9. Explain why U.S. environmental laws have been under attack since 1980.
10. Explain the importance of environmental security, relative to economic and military security.
11. Describe some of the harmful environmental impacts of war.
12. What are four guidelines for shifting to more environmentally sustainable societies?

Module 25 Ethics of Sustainability

Module 25 Overview

Major environmental worldviews differ on which is more important; human needs, or the overall health of ecosystems and the biosphere.

People disagree on how serious various environmental problems are and on what we should do about them. These conflicts arise out of differing environmental worldviews, how people think the world works and what they believe their role in the world should be. Part of that perceived role is determined by what one believes about what is right and what is wrong in our behavior toward the environment. Some environmental worldviews are human-centered, focusing on the needs and wants of people; others are life-centered, focusing on individual species, or the entire biosphere.

The first step to living more sustainably is to become environmentally literate. There is overwhelming evidence that humans are in the process of degrading our own life-support system and that during this century our behavior threatens not only human civilization, but the existence of up to half of the world's species. Part of the problem comes from ignorance about how the earth works, what we are doing to its life-sustaining systems, and how we can change our behavior toward the earth.

Many analysts say that formal environmental education is important, but not enough. They sway we must appreciate not just the economic value of nature, but also its ecological, aesthetic, and spiritual values. The problem is not just a lack of environmental literacy but also a lack of intimacy with nature and an incomplete understanding of how it sustains us. Which means that we face a dangerous paradox. At a time when humans have more power than ever before to disrupt nature, most people know little about, and have little direct contact with, nature.

Living sustainably is about living well, but within certain limits of consumption. Living sustainably has to be based on respect for the natural processes that keep us alive and support our economies. The goal is to protect the earth's precious biodiversity and provide a high quality of life for all life forms. Sustainability is not only about sustaining resources for us, it's also about sustaining the entire web of life, because all past, present, and future forms of life are connected.

Living more sustainably will not be easy. We wont make this transition by relying on technological fixes such as recycling, changing from incandescent to compact fluorescent lightbulbs, and driving energy-efficient hybrid cars. Effective long-term solutions have to start with this question: How do we greatly reduce the amounts of matter and energy resources we use and our ecological footprints?

In the end, it comes down to what individuals do to make the earth a better place to live for future generations, for other species, and for the ecosystems that support us. This means becoming and staying environmentally informed, evaluating and reducing the environmentally harmful aspects of our own life-styles, and becoming politically involved. We can vote with both our ballots and our pocketbooks to influence governments and businesses to help steer our societies onto a more sustainable path.

Module 25 Goals

The following is a list of goals that should be met by the end of this module. The goals are very broad and, sometimes, general in nature. This was done to allow you to decide how much depth you'll want to go into for each goal.

Examining the Progress Check questions will also help focus your attention better on the goals.

1. Identify major environmental worldviews.
2. Explain the role of education in living more sustainably.
3. Discuss how we can live more sustainably.

Module 25 Key Words

Here's a couple of key words for this module:

ethics
stewardship

Module 25 Progress Check

To check your understanding of the goals of this module, you should be able to answer the following questions.

1. What is the definition of environmental worldview?
2. What is meant by environmental ethics?
3. Define the following environmental worldviews: planetary management, stewardship, environmental wisdom, and deep ecology.
4. List three issues involved in deciding which species to protect from premature extinction as a result of our activities.
5. Discuss the controversy over whether we can effectively manage the earth.
6. List four questions that lie at the heart of environmental literacy.
7. Describe three ways in which we can learn from the earth.
8. Describe three traps that lead to denial, indifference, and inaction concerning the environmental problems we face.
9. List seven components of the environmental or sustainability revolution.
10. Describe connections between Biosphere 2, the transition to more environmentally sustainable societies, and the four scientific principles of sustainability.

FOLLOW-UP ACTIVITIES

The following activities are optional but are provided to apply you knowledge of environmental science to everyday activities. You can pick and choose the ones that are the most appropriate to you.

Module 3 - Ecosystems

Visit a nearby terrestrial ecosystem or aquatic life zone and try to identify major producers, primary and secondary consumers, detritus feeders, and decomposers.

Module 4 - Evolution and Biodiversity

Learn about the emerging field of synthetic biology, which combines biology, genetics, and engineering. How might synthetic biology get around some of the problems raised by genetic engineering and what problems does it pose?

Module 5 - Species Interactions & Populations

Identify the types of parasites that are likely to be found in our bodies, where they come from, their life cycles, and the damage they cause.

Module 6 - Human Population Study

Compare the age structure diagrams of the United States, Niger, and China. Use these diagrams to project their future population and economic growth as well as projected environmental and social problems they will face.

Module 7 - Biomes and Climate

Examine how vegetation and animal life in your area has changed over the past 50 years due to human activities.

Module 8 - Extinction

Identify a plant or animal that interests you and find out its numbers and global distribution, whether it is threatened with extinction, the future threats to its survival, and actions that are currently being taken to sustain it.

Module 9 - Maintaining Land Biomes

New Zealand has had a policy of meeting all its demand for wood, and wood products, by growing timber on managed tree plantations. Investigate this policy and evaluate the advantages and disadvantages of this approach in terms of its effectiveness.

Module 10 - Aquatic Biodiversity

Developers want to drain a large area of inland wetland in your community and build a large housing development. List the main arguments the developers would use to support this project and the main arguments ecologists would use in opposing it. Imagine you were an elected city official in this community, what arguments would you use for vote on this project either way.

Module 11 - Maintaining Aquatic Biodiversity

Pick a major seafood species and describe its life and reproductive cycles. Investigate if the species has been overfished, and if so, where in the world it has been depleted, by how much, and by what methods. Construct an argument for removing this species from a seafood restaurant's menu.

Module 12 - Food, Soil, and Pests

The ban of DDT has been criticized by many scientists as not being based solely on scientific investigation. Many claim that Rachel Carson in her book, *Silent Spring*, did not present enough peer reviewed evidence to warrant the worldwide ban. Critics of the ban claim that DDT nearly eliminated malaria and could have saved over 500,000 lives a year. Investigate this controversy and determine if the decision to ban DDT was correct or not.

Module 13 - Fresh Water Resources

Determine who owns and manages the supply and distribution of water in your community. Obtain the most recent water quality standards and compare them to the water quality of a neighboring community not supplied by your source.

Module 14 - Geology and Minerals

Research how a class ring, an LCD TV, and a car is made and how much material goes into manufacturing those items.

Module 15 - Nonrenewable Energy

Investigate the accident at the Three Mile Island (TMI) nuclear power plant near Harrisburg, Pennsylvania, in 1979. According to the nuclear power industry, the accident showed that its safety systems work because the accident caused no known deaths. Critics argue that the accident was a wake-up call about the potential dangers of nuclear power plants and that it led to tighter and better safety regulations. Provide evidence to support one of the above positions.

Module 16 - Renewable Energy

Make a list of the major appliances in your home (washer, dryer, stove, refrigerator, dish washer, air conditioner/heater, and TVs). For each appliance record the brand name and how much energy it uses every year. Investigate two competing brands and determine which ones are the most energy efficient. If you were to buy all of the most energy efficient appliances on your list, how much energy would you save every year?

Module 17 - Environmental Hazards

Identify an emerging viral disease and investigate how it spreads, its effect on the human body, possible treatments, and the strategies used for controlling its spread.

Module 18 - Solid and Hazardous Waste

What percentage of the municipal solid waste (MSW) in your community is (a) placed in a landfill, (b) incinerated, (c) composted, and (d) recycled? Investigate technologies used in local landfills and incinerators. Determine what leakage and pollution problems have local landfills or incinerators had. Does your community have a recycling program and is it voluntary or mandatory? Determine how your community benefits economically from the recycling program. Does your community have a hazardous waste collection program? What does your community do with the hazardous waste?

Module 19 - Water Pollution

Are the storm drains and sanitary sewers in your community combined or separate? Determine how your community handles these wastes. Investigate the kind of sewage treatment process your community uses and the plans your community has to upgrade the treatment plants.

Module 20 - Air Pollution

Identify the source(s) of air pollution in your community and compare the air quality of your community from one that is 50 miles away. Determine your community controls air pollution.

Module 21 - Climate Change

Calculate how much CO_2 your family car emits each day by taking the number of miles driven that day and multiplying it by 20. Divide the result by the number of miles per gallon your car gets. Add 20% to include the CO_2 emitted in the manufacturing of the gasoline you used. If your family has more than one car, determine which one has the lowest carbon footprint.

Module 22 - Environment and Cities

Investigate the land use policies of your community and the role that citizens play in these policies.

Module 23 - Economics of Sustainability

Choose a state or federal environmental regulation where you live (such as a water pollution law or regulation) and examine how it affects businesses and other organizations in your community. Talk to local businesses and determine if they view environmental regulations to be helpful to their business or not.

Module 24 - Politics of Sustainability

Choose a federal environmental law and contact your U.S. Congressman and U.S. Senators to determine their position on these laws. Find out if they think these laws should be strengthen, weaken, or eliminated. Ask them why they take their position.

Module 25 - Ethics of Sustainability

Answer the following fundamental ecological questions about the corner of the world you live in: Where does your water come from? Where does the energy you use come from? What kinds of soils are found in your community? What types of wildlife is there? Where does your food come from? Where does your waste go? Explain anything that was a total surprise to you.

WEB RESOURCES

The following list of web sites correspond to many of the modules in the guide. They are meant to provide some depth to the topics in the modules, and also as a springboard for further in-depth searches.

All these links worked at the time this page was published. But the very nature of the web means that some of these could disappear at any time. If you find any of them have stopped working, please email me so I can update the list.

Air Data - EPA

http://www.epa.gov/airdata

AIRNow - Air Quality Conditions & Forecasts

http://airnow.gov

Air Quality - NASA

http://earthobservatory.nasa.gov/Experiments/CitizenScientist/

AirQuality/

Biomes of the World

https://php.radford.edu/~swoodwar/biomes/

Climate Change - EPA

http://www/epa.gov/climatechange/

The Donora Fluoride Fog

http://www.fluoridation.com/donora.htm

Dr. Blythe's Rainforest Education

http://www.rainforesteducation.com/

Ecological Footprint Quiz

http://www.myfootprint.org/

Environmental Case Studies

http://arcticcircle.uconn.edu/VirtualClassroom/menu.html

Environmental Disasters

http://en.wikipedia.org/wiki/List_of_environmental_disasters

Environmental Issues

http://en.wikipedia.org/wiki/List_of_environmental_issues

The Environmental Literacy Council

http://www.enviroliteracy.org/

Environmentalism

http://en.wikipedia.org/wiki/Environmentalism

Exploring the Environment

http://ete.cet.edu/modules/modules.html

The Habitable Planet

http://www.learner.org/courses/envsci/index.html

Inspiration Green

http://www.inspirationgreen.com/food-organic-choices

NASA Global Maps

http://earthobservatory.nasa.gov/GlobalMaps/

Our Ocean Planet

http://oceanworld.tamu.edu/resources/oceanography-book/contents.htm

Precipitation

http://earthobservatory.nasa.gov/Experiments/CitizenScientist/Precipitation/

Rainforest School Reports

http://www.rain-tree.com/schoolreports.htm

Recycling Numbers

http://pslc.ws/macrog/work/recycle.htm
River of Venom
http://www.accessexcellence.org/AE/mspot/rov/
Science Daily: Environmental Science News
http://www.sciencedaily.com/news/earth_climate/environmental_science/
Sustainability
http://en.wikipedia.org/wiki/Sustainability
Toxics Release Report
http://www2.epa.gov/toxics-release-inventory-tri-program
U.S. Geologic Survey
http://www.usgs.gov/climate_landuse/
The Water Cycle
http://ga.water.usgs.gov/edu/watercycle.html
Water Quality
http://earthobservatory.nasa.gov/Experiments/CitizenScientist/WaterQuality/
The World Factbook
https://www.cia.gov/library/publications/the-world-factbook/geos/ng.html
WHO Global Health Atlas
http://apps.who.int/globalatlas/
World in The Balance
http://www.pbs.org/wgbh/nova/worldbalance/

ENVIRONMENTAL ORGANIZATIONS

All these links all worked at the time this page was published. But the very nature of the web means that some of these could disappear at any time. If you find that any of them have stopped working, please email me so I can update this list.

Air Now

www.airnow.gov

American Farmland Trust

www.farmland.org

American Lung Association

www.lungusa.org

American Rivers, Inc.

www.amrivers.org

Bread For The World

www.bread.org

Center For Science In The Public Interest

www.cspinet.org

Center For Climate And Energy Solutions

www.c2es.org

Chesapeake Bay Foundation

www.cbf.org

Clean Water Action Project

www.cleanwateraction.org

Common Cause

www.commoncause.org

Community Transportation Association of America

www.ctaa.org

Congress Watch

www.citizen.org/congress

Conservation International

www.conservation.org

Coral Reef Alliance

www.coral.org

Critical Mass Energy & Environmental Project

www.citizen.org/cmep

Defenders of Wildlife

www.defenders.org

Ducks Unlimited

www.ducks.org

Earthwatch Institute

www.earthwatch.org

Environmental Defense Fund

www.edf.org

Environmental Law Institute

www.eli.org

Environmental Literacy Council

www.enviroliteracy.org

Forest Stewardship Council

www.fscus.org

Freedom From Hunger

www.freefromhunger.org

Friends Of The Earth

www.foe.org

Global Action Plan

http://empowermentinstitute.net

Greenpeace USA, Inc

www.greenpeace.org

Habitat For Humanity International

www.habitat.org

Heifer Project International

www.heifer.org

Institute For Local Self-Reliance

www.ilsr.org

International Food Policy Research Institute

www.ifpri.org

Izaak Walton League of America, Inc

www.iwla.org

Land Trust Alliance

www.lta.org

League Of Conservation Voters

www.lcv.org

League Of Women Voters Of The U.S.

www.lwv.org

Millennium Ecosystem Assessment

www.millenniumassessment.org

National Audubon Society

www.audubon.org

National Council For Science And The Environment

www.ncseonline.org

National Park Foundation

www.nationalparks.org

National Parks Conservation Association

www.npca.org

National Wildlife Federation

www.nwf.org

National Resources Defense Council

www.nrdc.org

Nature Conservancy

http://nature.org

North American Association For Environmental Education

www.naaee.net

Oxfam America

www.oxfamamerica.org

Pew Center On Global Climate Change

www.pewclimate.org

Planned Parenthood Federation Of America

www.plannedparenthood.org

The Population Institute

www.populationinstitute.org

Population Connection

www.populationconnection.org

Population Reference Bureau, Inc.

www.prb.org

Rachel Carson Council, Inc.

http://members.aol.com/rccouncil/our-page/rcc_page.htm

Rain Forest Action Network

www.ran.org

Rainforest Alliance

www.rainforestalliance.org

Resources For The Future

www.rff.org

Science Daily

http://www.sciencedaily.com/news/earth_climate/environmental_science/

Shack/Slum Dwellers International

www.sdinet.org

Sierra Club

www.sierraclub.org

Sprawl Watch Clearinghouse

www.sprawlwatch.org

Trust For Public Land

www.tpl.org

Union Of Concerned Scientists

www.ucsusa.org

U.S. Environmental Protection Agency

www.epa.gov

U.S. Public Interest Research Group

www.pirg.org

Water Environment Federation

www.wef.org

Wilderness Society

www.tws.org

World Resources Institute

www.wri.org

World Wildlife Fund

www.worldwildlife.org

World Watch Institute

www.worldwatch.org

EXTENDING THIS COURSE

I offer a free newsletter full of tips on homeschooling science, and have both a weblog with science articles, links, recommended resources, and an online forum.

http://sciencelessonsforkids.com/sign-up

I am also developing course materials that integrate not only with this guide, but also with the Week-by-Week Basic Biology Guide and the Advanced Biology: Homeschooling Week By Week guide.

Look for these modules at **http://sciencelessonsforkids.com**.

Made in the USA
Middletown, DE
19 October 2022

13119711R00113